Math Kangaroo Competition
Problems and Solutions
Grades 5 and 6
Volume II

Problems and Solutions
Odd Years
1999–2021

Editor in Chief
Agata Gazal
Chief Editorial Officer for Math Kangaroo
Bozeman, MT

Reviewers
Joanna Matthiesen
Chief Executive Officer for Math Kangaroo
Granger, IN

Izabela Szpiech
Chief Financial Officer for Math Kangaroo
Norridge, IL

Kasia Nalaskowska
Chief Information Officer for Math Kangaroo
Aurora, IL

Svetlana Savova
Chief Academic Officer for Math Kangaroo
Johns Creek, GA

Contributors
Maria Omelanczuk
Former CEO and President for Math Kangaroo
Oswego, IL

Andrzej Zarach, PhD
Math Content Reviewer, Professor Emeritus of East Stroudsburg University
East Stroudsburg, PA

Dawid Zarach
Math Content Reviewer
East Stroudsburg, PA

Cover and Graphics Credit
Magdalena Teodorowicz
Chief Design Officer for Math Kangaroo
Cordova, TN

Agata Gazal
Chief Editorial Officer for Math Kangaroo
Bozeman, MT

We would like to give special thanks to other countless people who contributed to the questions and solutions of this book since 1998, chiefly to the Math Kangaroo question writers from all over the world that are part of the AKSF organization (www.aksf.org); Math Kangaroo solution writers also include Math Kangaroo USA competition organizers and Math Kangaroo Alumni (www.mathkangaroo.org). We would also like to thank the hundreds of educators who gave us feedback on the questions and solutions, and finally the tens of thousands of students that take the challenge each year. Thank you all for your help in developing this book.

© Copyright 2021 by Math Kangaroo in USA, NFP, Inc.
www.mathkangaroo.org

Printed by:
Classic Printing & Thermography
Wood Dale, IL

For additional copies of this book, please contact the publisher:
Math Kangaroo USA
info@mathkangaroo.org

ISBN 978-0-578-98454-4

Preface

Many people enjoy the challenge of solving math riddles and other types of puzzles. This book presents 360 entertaining problems and solutions presented to 5th and 6th grade students during the Math Kangaroo Competition odd years spanning 1999-2021, the total of 12 tests. Each test consists of 30 questions divided into easy, medium, and difficult categories. The questions were selected at the annual *Kangourou sans Frontières* meeting where mathematicians from over 80 countries work together to choose the most engaging and age-appropriate questions for the annual Math Kangaroo competition.

This easy-to-use resource book includes fun questions, pictures, and interesting solutions that challenge children to use math and logic as a tool for understanding the world around them. Problem solving is a skill that all children use, sometimes without even knowing it. This book will help students practice their math skills that often involve logical reasoning and reflecting on the solutions.

We hope this book will be cherished not just by students who love mathematics but also by teachers who are passionate about teaching unconventional and challenging math. We believe students will benefit from this book and find it both insightful and entertaining.

Joanna Matthiesen

President and CEO of Math Kangaroo USA

Table of Contents

Part I: Problems 7
    Problems from Year 1999 9
    Problems from Year 2001 13
    Problems from Year 2003 18
    Problems from Year 2005 22
    Problems from Year 2007 27
    Problems from Year 2009 31
    Problems from Year 2011 36
    Problems from Year 2013 41
    Problems from Year 2015 46
    Problems from Year 2017 50
    Problems from Year 2019 55
    Problems from Year 2021 60

Part II: Solutions 67
    Solutions for Year 1999 69
    Solutions for Year 2001 75
    Solutions for Year 2003 81
    Solutions for Year 2005 87
    Solutions for Year 2007 92
    Solutions for Year 2009 99
    Solutions for Year 2011 105
    Solutions for Year 2013 111
    Solutions for Year 2015 119
    Solutions for Year 2017 126
    Solutions for Year 2019 133
    Solutions for Year 2021 141

Part III: Answer Keys 151

Part I: Problems

Problems from Year 1999

Problems 3 points each

1. $1999 - 999 + 99 - 9 =$

   (A) 1900 (B) 1090 (C) 1000 (D) 1990 (E) 1009

2. Anna and her sister Mary take two different ways to school. Whose way is longer?

   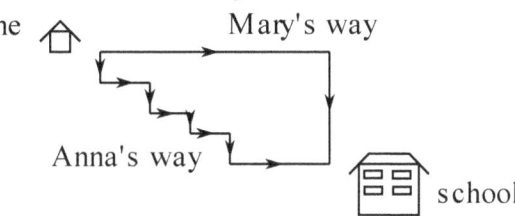

   (A) Anna's way (B) Mary's way (C) It depends on the distance to the school.
   (D) Both ways have the same length. (E) It is impossible to determine.

3. One quarter of one half of double 32 is equal to

   (A) 4 (B) 8 (C) 16 (D) 32 (E) 64

4. How many different ways connect city A to city C?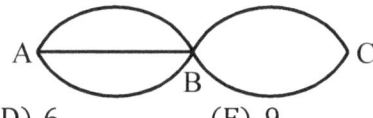

   (A) 2 (B) 3 (C) 5 (D) 6 (E) 9

5. How many times faster does the minute hand of any clock move than the hour hand?

   (A) 6 (B) 12 (C) 9 (D) 10 (E) 15

6. In the picture to the right some small squares are shaded. The big square has dimensions of 9×9 and is made up of both shaded and white small squares. Find the difference between the number of shaded squares and white squares.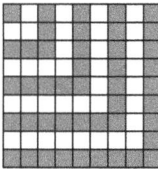

   (A) 0 (B) 1 (C) 5 (D) 9 (E) 10

7. ABCD is a square and point M is the midpoint of side AB. The shaded part has an area equal to 9 cm² (see the picture). What is the area of square ABCD?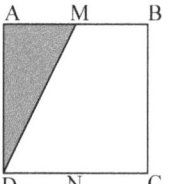

   (A) 18 cm² (B) 27 cm² (C) 32 cm² (D) 36 cm² (E) 45 cm²

PROBLEMS 1999

8. A movie started at 1:47 p.m. and finished at 4:18 p.m. How long was the movie?

   (A) 185 min (B) 151 min (C) 91 min (D) 149 min (E) 209 min

9. At a birthday party, one person found out that no two of the people present were born in the same month. What is the greatest possible number of people at that party?

   (A) 11 (B) 12 (C) 13 (D) 24 (E) 44

10. How many bricks were taken out from the wall shown in the picture?

    (A) 26 (B) 32 (C) 36 (D) 40 (E) 42

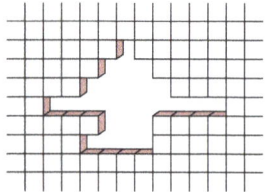

Problems 4 points each

11. John opened a book and found that the sum of the page number on the left and the page number on the right is equal to 25. What is the product of the two page numbers?

    (A) 169 (B) 144 (C) 150 (D) 156 (E) 1998

12. A certain kangaroo's jump is 5 m long. How many jumps does it need to make to cover a distance of 5000 m + 5000 dm + 5000 cm + 5000 mm?

    (A) 1000 (B) 1100 (C) 1110 (D) 1111 (E) 5555

13. A stick that in reality measures 1 m is 2 cm long in a certain picture, and in that same picture the height of a fence is 4.5 cm. What is the actual height of the fence in cm?

    (A) 450 (B) 225 (C) 45 (D) 22.5 (E) 4.5

14. Find the sum of the missing digits.

    (A) 6 (B) 8 (C) 10 (D) 12 (E) 14

15. Which of the figures below will form a rectangle when combined with the figure to the right?

    (A)  (B)  (C)  (D)  (E)

16. The dog is 9 times as heavy as the cat, the mouse is 20 times lighter than the cat, and the turnip is 6 times as heavy as the mouse. How many times is the dog as heavy as the turnip?

(A) 30   (B) 27   (C) 1080   (D) 15
(E) The dog is lighter than the turnip.

17. A kangaroo wants to make a rectangular quilt 1.5 m long and 1 m wide using square scraps which measure 10 cm × 10 cm. At every point where four squares meet she wants to place a fancy button. How many buttons will she need?

(A) 150   (B) 104   (C) 126   (D) 140   (E) 135

18. One bowl contained 26 liters of water and another bowl contained 7 liters of water. The same amount of water was added to each bowl, and now the second bowl contains 3 times less water than the first bowl. How many liters of water were added to each bowl?

(A) 2.5   (B) 5   (C) 7.5   (D) 10   (E) 15

19. In a magic square, the sum of the numbers in the cells in each vertical column, in each horizontal row, and along each diagonal is the same. The square shown to the right is a magic square. Two of the numbers have been taken out, and three were covered with the letters A, B, and C. Find the sum A + B + C.

16	3	A
C	10	
B		4

(A) 30   (B) 41   (C) 14   (D) 25
(E) It is impossible to determine.

20. John and Adam are making a square using small squares with the same dimensions. Adam places one red square down, and John adds 8 green squares around it to make a second square. Then Adam places 16 yellow squares around this square to make a third square, and so forth. How many squares does Adam have to place to make the fifth square?

(A) 32   (B) 64   (C) 81   (D) 121   (E) 125

Problems 5 points each

21. Ela came to Anna's birthday party 5 minutes earlier than Stan but 3 minutes later than Iwona. Iwona left first. She left 2 minutes earlier than Stan and 5 minutes earlier than Ela. How many minutes longer was Ela at the party than Stan?

(A) 2   (B) 4   (C) 6   (D) 8   (E) Stan stayed longer than Ela.

PROBLEMS 1999

22. Six hundred and six people can eat six hundred and six hotdogs, six hundred of which are with mustard and six are without mustard. How many hotdogs without mustard will you need for six hundred and six thousand six hundred and six people?

    (A) 606   (B) 1000   (C) 6006   (D) 606,606   (E) 600,600

23. Four squirrels ate 1999 nuts altogether, and each one ate at least 100 nuts. The first squirrel ate more nuts than any other squirrel. The second and third squirrels together ate 1265 nuts. How many nuts did the first squirrel eat?

    (A) 598   (B) 271   (C) 629   (D) 634   (E) other answer

24. When it is raining, the cat stays in the room or in the basement. When the cat stays in the room, the mouse is in the foyer and the cheese is in the refrigerator. When the cheese is on the table and the cat stays in the basement, the mouse is in the room. Right now, it is raining and the cheese is on the table. So, for sure:

    (A) The cat is in the room.
    (B) The cat is in the room and the mouse is in the foyer.
    (C) The mouse is in the foyer.
    (D) The cat is in the basement and the mouse is in the room.
    (E) This situation is impossible.

25. What is the maximum number of acute angles that can be made by six rays coming out from the same point?

    (A) 6   (B) 8   (C) 9   (D) 12   (E) 15

26. The "yield" of 36 equals 18, the "yield" of 325 equals 30, the "yield" of 45 equals 20, and the "yield" of 30 equals 0. What is the "yield" of 531?

    (A) 10   (B) 15   (C) 16   (D) 21   (E) 22

27. The diagram to the right corresponds to only one of the cubes below. Which one?

PROBLEMS 1999

28. In each one of the 5 cups shown below you can find coffee, cocoa, or tea. There is twice as much coffee as cocoa. None of the beverages was poured into three cups. In which cup is the cocoa?

(A) 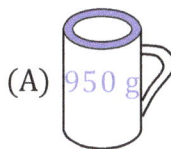 950 g (B) 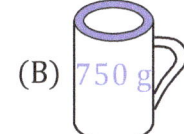 750 g (C) 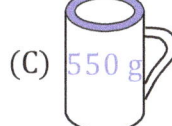 550 g (D) 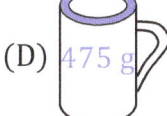 475 g (E) 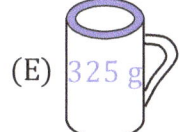 325 g

29. How many squares are there in the figure to the right?

(A) 46 (B) 47 (C) 45 (D) 33 (E) 37

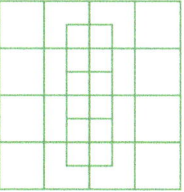

30. An electronic clock shows hours, minutes, and seconds. Right now it shows 19:58:47. As you can see, all the digits are different. After how many seconds will you see all different digits again?

(A) 40 (B) 73 (C) 156 (D) 157 (E) 898

Problems from Year 2001

Problems 3 points each

1. In the year 2001 the Little Kangaroo calculated that the value of the expression $2 \times 0 + 0 \times 1$ is

(A) 2 (B) 0 (C) 1 (D) 2001 (E) 3

2. What part of the rectangle does the shaded region represent? (See the picture.)

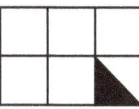

(A) $\frac{1}{6}$ (B) $\frac{1}{8}$ (C) $\frac{1}{10}$ (D) $\frac{1}{12}$ (E) $\frac{1}{15}$

3. Which of the 5 napkins below was made from the cut-out you can see on the right?

(A) 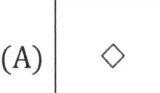 (B) 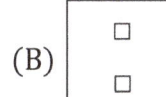 (C)  (D) 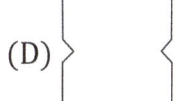 (E)

4. At noon grandpa set his clock to the correct time. The clock is running 20 seconds late per hour. After 24 hours, the clock will be late:

(A) 7 minutes   (B) 8 minutes   (C) 9 minutes   (D) 10 minutes   (E) 11 minutes

5. A certain plane can carry 108 passengers. During one of its flights, Anita noticed that not all the seats were occupied. There were twice as many occupied seats as empty ones. How many passengers were on the plane?

(A) 36   (B) 42   (C) 56   (D) 64   (E) 72

6. Johnny has 3 sisters and 5 brothers. His sister Ella found that the product of the number of her sisters and the number of her brothers is:

(A) 8   (B) 10   (C) 12   (D) 15   (E) 18

7. Simon thought of a certain number. Then, he made the number twice as big. He then doubled the result, and then doubled it again, and then doubled it once more. Which of the numbers below cannot be the result of this operation?

(A) 80   (B) 1200   (C) 48   (D) 84   (E) 880

8. Which of the shaded regions has the largest area?

9. Picture 1 shows a way of writing 14, and Picture 2 shows a way of writing 123. What number does Picture 3 represent?

Picture 1   Picture 2   Picture 3

(A) 1246   (B) 2461   (C) 2641   (D) 1462   (E) other answer

10. What is the smallest number of matches you have to add to the puzzle shown in the picture in order to have 11 squares?

(A) 2   (B) 3   (C) 4   (D) 5   (E) 6

Problems 4 points each

11. Peter runs a lap around the track in 3 minutes and Paul in 4 minutes. Peter and Paul start to run around the track at the same time. After how many minutes will the boys meet at the starting point again?

(A) 6 (B) 8 (C) 10 (D) 12
(E) It depends on the distance around the track.

12. Anya has 201 coins. One third of her coins are 1 dollar coins. The same number of her coins are 2 dollar coins. The rest are 5 dollar coins. How much money does Anya have?

(A) 536 dollars (B) 201 dollars (C) 516 dollars (D) 1020 dollars (E) 2001 dollars

13. At the Olympics there was a 10 kilometer race. Adam did not finish the race and asked his friends the following question: "How many meters did I have left to finish the race if I already ran 8,631 meters, 3,456 decimeters, and 12,340 centimeters?"
(Note:1 kilometer is 1,000 meters, 1 meter is 10 decimeters, and 1 meter is 100 centimeters.)
The answer to his question is:

(A) 1060 (B) 160 (C) 900 (D) 100 (E) 1000

14. There are 7 sticks, each 14 centimeters long. They were put together in the way shown in the picture, for a total length of 80 centimeters. Each of the segments marked with a question mark is the same length. How long are these segments?

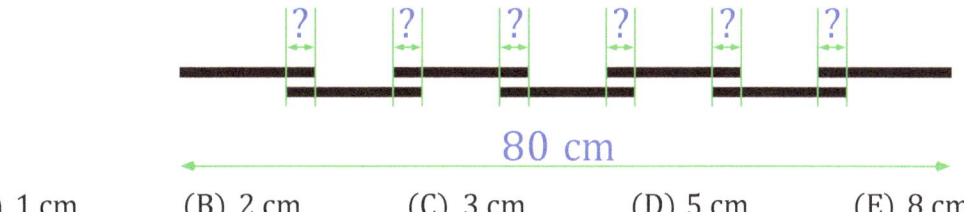

(A) 1 cm (B) 2 cm (C) 3 cm (D) 5 cm (E) 8 cm

15. If the red dragon had 6 heads more than the green dragon, then both dragons would together have 34 heads. Actually, the red dragon has 6 heads fewer than the green dragon. How many heads does the red dragon have?

(A) 6 (B) 8 (C) 12 (D) 14 (E) 16

16. The length of a certain rectangle is 80 cm and its area is 3,200 cm². Find the length of another rectangle if its area and width are half the area and the width of the rectangle described above.

(A) 20 cm (B) 40 cm (C) 60 cm (D) 80 cm (E) 100 cm

PROBLEMS 2001

17. Zosia spends 1 hour doing her homework. She spends one third of this time doing math, and she spends two fifths of the remaining time working on geography. How many minutes does she spend doing the rest of her homework?

   (A) 12   (B) 20   (C) 24   (D) 36   (E) 40

18. The triplets Adam, Ian, and Stan and their sister Mary, who is four years older than they are, altogether were 24 years old three years ago. How old is Mary today?

   (A) 5   (B) 8   (C) 9   (D) 12   (E) 15

19. Jonah's garden is shaped in the way shown in the picture. The lengths of its sides are given in meters, and any two adjacent sides are perpendicular. What is the area of the garden in square meters?

   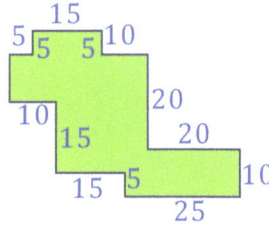

   (A) 700   (B) 750   (C) 800   (D) 850   (E) 900

20. Together, Adam, Bart, and Charlie earned 280 dollars during their vacation. Adam made twice as much money as Bart and four times as much as Charlie. How many dollars did Charlie earn?

   (A) 30   (B) 40   (C) 50   (D) 60   (E) 70

Problems 5 points each

21. 9 is the ones digit in the product of 7 × 7. 3 is the ones digit in the product of 7 × 7 × 7. What is the ones digit in the number which is the product of 100 sevens?

   (A) 1   (B) 3   (C) 5   (D) 7   (E) 9

22. The biggest attraction of the Fair is the Big Wheel. The picture shows an example of such a ride but with fewer cabins. The cabins are equally spaced and are numbered consecutively with numbers 1, 2, 3, and so on. When the cabin with the number 25 is at the very top, the cabin with the number 8 is at the very bottom. How many cabins does the Big Wheel have?

   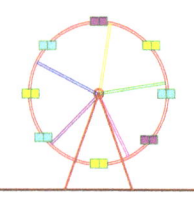

   (A) 33   (B) 34   (C) 35   (D) 36   (E) 37

PROBLEMS 2001

23. Each of the solids below is made of the same number of identical little cubes. For which of these solids would you need the largest amount of paint to paint its entire surface?

   (A) 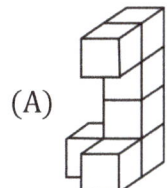 (B) 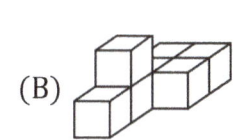 (C) 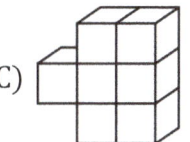 (D) 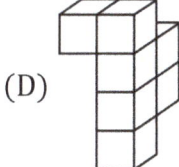 (E)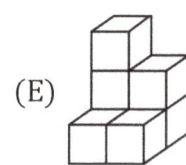

24. The area of the biggest square shown in the picture to the right is 16 cm² and the area of the smallest square is 4 cm². What is the area of the medium square?

   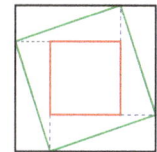

   (A) 8 cm²   (B) 8½ cm²   (C) 10 cm²   (D) 10½ cm²   (E) 12 cm²

25. The sum of the number of dots on any two opposite sides of any die is 7. Jo is building a solid using six dice which she is gluing together in the way shown in the picture. What is the largest possible number of dots that she can get on the surface of the solid?

   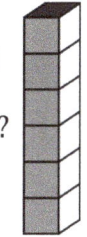

   (A) 106   (B) 91   (C) 95   (D) 84   (E) 96

26. Fill in the blanks with digits so that the equality is true: **45 × _3 = 3_ _ _**. The sum of the four digits placed in the blanks is:

   (A) equal to 20   (B) equal to 21   (C) equal to 17
   (D) more than 21   (E) less than 17

27. Inside the big cube, which was made out of small cubes, tunnels were made going through the cube in such a way that they are parallel to its walls (see the picture). After making the tunnels, how many small cubes are left in the solid?

   (A) 88   (B) 80   (C) 70   (D) 96   (E) 85

28. A one-hundred-year-old oak generates 1.7 kg of oxygen per hour. How many of these trees do we need to supply oxygen to 34 students for 1 hour if we know that each student consumes 0.7 kg of oxygen per hour by breathing?

   (A) 10   (B) 12   (C) 14   (D) 15   (E) 21

29. The vertices of the star shown in the picture are the midpoints of the sides of a regular hexagon. If the area of the star is 6 then the area of the hexagon is:

   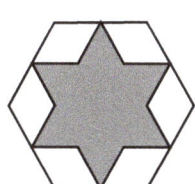

   (A) 8   (B) 9   (C) 12   (D) 15   (E) 18

30. Using digits from 1 to 6, each digit only once, we form two 3-digit numbers, for instance, 645 and 321, and then we subtract the smaller number from the larger one. In this example, the difference is 324. What is the smallest possible value of the difference you can get this way?

    (A) 69 (B) 56 (C) 111 (D) 47 (E) 38

Problems from Year 2003

Problems 3 points each

1. Which of the following numbers is the greatest?

    (A) $2 + 0 + 0 + 3$ (B) $2 \times 0 \times 0 \times 3$ (C) $(2 + 0) \times (0 + 3)$
    (D) $20 \times 0 \times 3$ (E) $(2 \times 0) + (0 \times 3)$

2. Zosia is drawing flowers of different colors. The first flower is blue, the second white, the next one red, the next one yellow, and again blue, white, red, yellow, and so on in the same order. What is the color of the twenty ninth flower drawn by Zosia?

    (A) blue (B) white (C) red (D) pink (E) yellow

3. How many integers are there on the number line between the numbers 2.09 and 15.3?

    (A) 13 (B) 14 (C) 11 (D) 12 (E) Infinitely many

4. The least positive integer which is divisible by 2, 3, and 4 is:

    (A) 1 (B) 6 (C) 12 (D) 24 (E) 36

5. Two of the numbers located on the two circles (see the picture) are represented by letters A and B. The sum of the numbers on each circle is equal to 55. What number is represented by the letter A?

    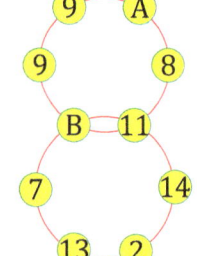

    (A) 9 (B) 10 (C) 13 (D) 16 (E) 17

6. Tom has 9 bills worth 100 dollars each, 9 bills worth 10 dollars each, and 10 coins worth 1 dollar each. How much money does Tom have?

    (A) $1,000 (B) $991 (C) $9,910 (D) $9,901 (E) $99,010

PROBLEMS 2003

7. A square with the length of its side equal to x is made up of a square with an area of 81 cm², two rectangles with areas of 18 cm² each, and a small square. What is the value of x?

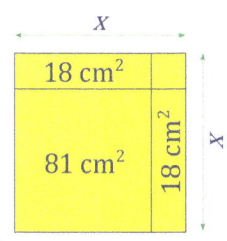

(A) 2 cm (B) 7 cm (C) 9 cm (D) 10 cm (E) 11 cm

8. The value of the expression $\frac{2003+2003+2003+2003+2003}{2003+2003}$ is equal to:

(A) 2003 (B) $\frac{1}{3}$ (C) 3 (D) $\frac{5}{2}$ (E) 6009

9. Dasia likes to add the digits that indicate the current time on her electronic watch. For example, when the watch shows 21:17, she gets a sum equal to 11. What is the greatest sum she can get? (Hint: in some countries and sometimes in the USA, instead of saying, "It is 1 p.m.," people may say, "It is 13:00." When it is 2 p.m., they may say, "It is 14:00," and when it is 12 a.m., they may say, "It is 24:00." In this problem 21:17 means 9:17 p.m. Time expressed using this method is sometimes called *military time*.)

(A) 24 (B) 36 (C) 19 (D) 25 (E) 28

10. The picture shows Clown Ian dancing on top of two balls and a cube. The radius of the lower ball is 6 dm, and the radius of the upper ball is three times shorter. The edge of the cube is 4 dm longer than the radius of the upper ball. At what height is Ian dancing?

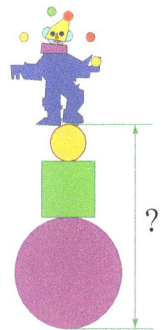

(A) 14 dm (B) 20 dm (C) 22 dm (D) 24 dm (E) 28 dm

Problems 4 points each

11. Let $AC = 10$ m, $BD = 15$ m, $AD = 22$ m (see the picture below). The length of segment BC is equal to

(A) 1 m (B) 2 m (C) 3 m (D) 4 m (E) 5 m

12. How many shortest distances along the edges of the cube are there that connect vertex A with the opposite vertex B?

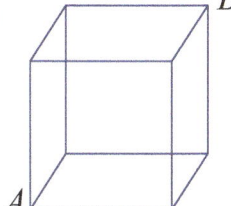

(A) 4 (B) 6 (C) 3 (D) 12 (E) 16

© Math Kangaroo in USA, NFP

13. Two pieces are cut out from a square puzzle. These two pieces made up the shaded region (see the picture). From the four pictures below, which two pieces are they?

1 2 3 4

(A) 1 and 4 (B) 2 and 4 (C) 2 and 3 (D) 1 and 3 (E) 3 and 4

14. We add together two different numbers chosen from the numbers 1, 2, 3, 4, and 5. How many different sums can we get?

(A) 5 (B) 6 (C) 7 (D) 8 (E) 9

15. The figure in the picture consists of 7 squares. Square A has the greatest area and square B has the smallest area. The lengths of the sides of two squares are given. How many squares identical to square B would it take to fill in square A completely?

(A) 16 (B) 25 (C) 36 (D) 49
(E) It is impossible.

16. A certain bar code consists of 17 black bars. A white bar is placed between any two black bars. The first bar and the last bar in the code are black. There are two kinds of black bars: wide and narrow. The number of white bars is 3 more than the number of wide black bars. How many narrow black bars are there in this bar code?

(A) 1 (B) 2 (C) 3 (D) 4 (E) 5

17. Eva has 20 marbles of four colors: yellow, green, blue, and black. 17 of them are not green, 5 are black, and 12 are not yellow. How many blue marbles does Eva have?

(A) 3 (B) 4 (C) 6 (D) 7 (E) 8

18. There are 17 trees on one side of the street on Tom's way from his house to school. One day Tom marked these trees with white chalk in the following way: on the way from his house to school he marked every other tree, starting with the first one, and on his way back home he marked every third tree, starting with the first one. How many trees were not marked?

(A) 4 (B) 5 (C) 6 (D) 7 (E) 8

PROBLEMS 2003

19. Today's date is written as 3.21.2003 (March 21st, 2003) and the time is 20:03 (8:03 p.m.). How will the date be written after 2003 minutes?

(A) 3.21.2003 (B) 3.22.2003 (C) 3.23.2003 (D) 4.21.2003 (E) 4.22.2003

20. What is the ones digit in the number 2003^{2003}?

(A) 7 (B) 1 (C) 9 (D) 5 (E) 3

Problems 5 points each

21. With how many zeros does the product of the consecutive natural numbers from 1 to 50 end?

(A) 5 (B) 10 (C) 12 (D) 20 (E) 50

22. The square ABCD consists of a white square and four shaded rectangles. Each of the shaded rectangles has a perimeter of 40 cm. The area of square ABCD equals:

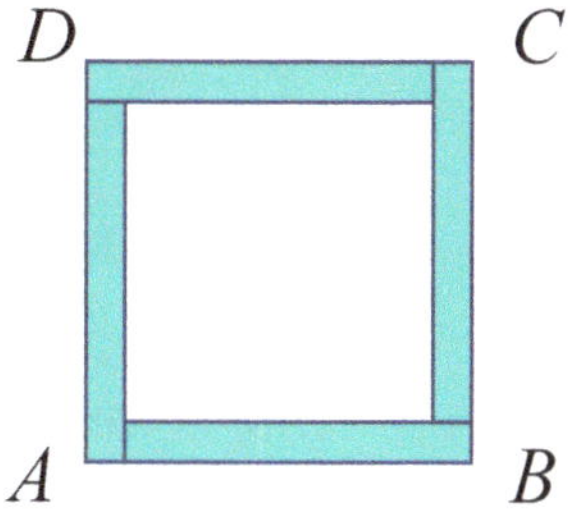

(A) 100 cm² (B) 200 cm² (C) 160 cm²
(D) 400 cm² (E) 80 cm²

23. We have six segments with lengths of 1, 2, 3, 2001, 2002, and 2003. In how many ways can we select three of these segments to build a triangle?

(A) 1 (B) 3 (C) 5 (D) 6 (E) 10

24. Pete is writing the numbers from 0 to 109 into a five-column table using a rule which is easy to figure out (see the picture to the right). Which of the pieces below cannot be filled in with numbers to fit Pete's table?

(A) 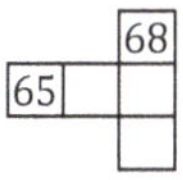 68/65 (B) 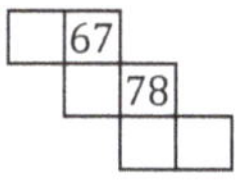 67/78 (C) 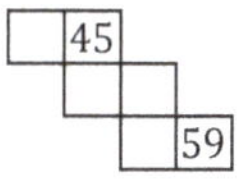 45/59 (D) 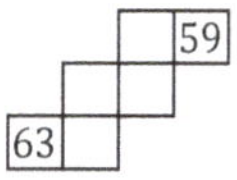 59/63 (E)  43/56

25. In the figure, the beginning part of the path from point A to point B is shown. How long is the whole path?

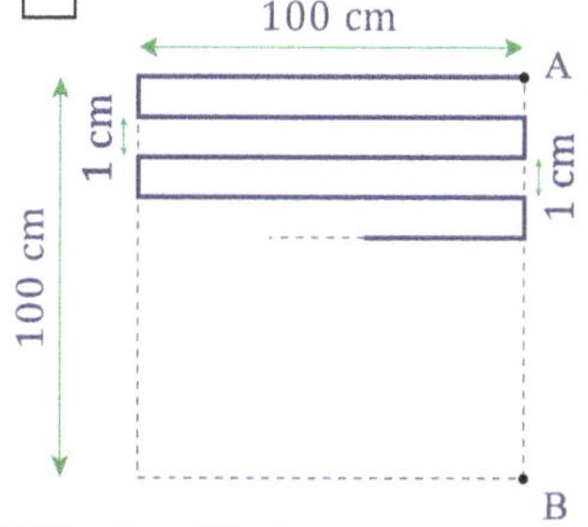

(A) 10,200 cm (B) 2,500 cm (C) 909 cm
(D) 10,100 cm (E) 9,900 cm

26. At 3:00 o'clock the minute hand and the hour hand make a right angle. What will the measure of the angle between these hands be after 10 minutes?

(A) 90° (B) 30° (C) 80° (D) 60° (E) 35°

© Math Kangaroo in USA, NFP   www.mathkangaroo.org

PROBLEMS 2003

27. In the addition problem shown in the picture, every square stands for a certain digit, every triangle stands for another digit, and every circle denotes yet another digit. What is the sum of the numbers represented by the square and the circle?

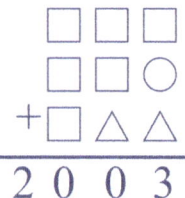

(A) 6   (B) 7   (C) 8   (D) 9   (E) 13

28. The shaded figure in the picture to the right consists of five identical isosceles right triangles. The area of the shaded figure is:

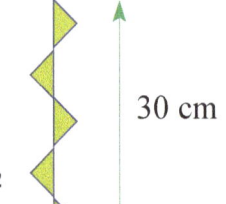

(A) 20 cm²   (B) 25 cm²   (C) 35 cm²   (D) 45 cm²   (E) 60 cm²

29. Red and green dragons lived in a cave. Each red dragon had 6 heads, 8 legs, and 2 tails. Each green dragon had 8 heads, 6 legs, and 4 tails. There were 44 tails altogether, and there were 6 fewer green legs than red heads. How many red dragons lived in the cave?

(A) 6   (B) 7   (C) 8   (D) 9   (E) 10

30. Anya has 9 crayons in a box. At least one of them is blue. At least two of every 4 crayons are of the same color, and at most three out of every 5 crayons are of the same color. How many blue crayons are in this box?

(A) 2   (B) 3   (C) 4   (D) 1   (E) 5

Problems from Year 2005

Problems 3 points each

1. A butterfly sat down on a correctly solved problem. What number did it cover up?

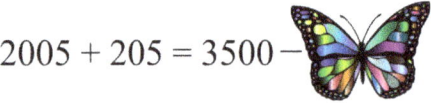

(A) 1295   (B) 1190   (C) 1390   (D) 1195   (E) 1290

2. Together, Anna and Olla have ten pieces of candy. Olla has two more pieces of candy than Anna. How many pieces of candy does Olla have?

(A) 8   (B) 7   (C) 6   (D) 5   (E) 4

PROBLEMS 2005

3. There are eight kangaroos in the table (see the picture). What is the least number of kangaroos that have to be moved to the empty cells of the table in order to have two kangaroos in each row and each column?

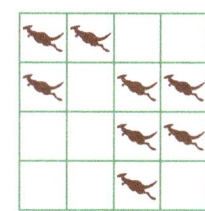

(A) 0 (B) 1 (C) 2 (D) 3 (E) 4

4. Eva lives with her parents, a brother, a dog, two cats, two parrots, and four goldfish. How many legs do they have altogether?

(A) 40 (B) 32 (C) 28 (D) 24 (E) 22

5. $2005 \times 100 + 2005 =$

(A) 22055 (B) 20052005 (C) 20072005
(D) 202505 (E) 2005002005

6. An ant is walking from point A to point B on a cube along the indicated path. The edge of the cube is 12 cm long. How far does the ant need to travel?

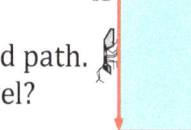

(A) 40 cm (B) 48 cm (C) 50 cm (D) 60 cm (E) 36 cm

7. On a shelf, there are 24 marbles in three colors: white, red, and brown. $\frac{1}{8}$ of the marbles are white and $\frac{2}{3}$ of the rest of the marbles are red. How many of the marbles are brown?

(A) 4 (B) 5 (C) 6 (D) 7 (E) 8

8. There are five cards on the table, labeled with numbers 1 to 5 as shown in the top row. One move consists of switching any two cards. What is the smallest number of moves needed so that the cards are arranged in the way shown in the bottom row?

(A) 2 (B) 4 (C) 1 (D) 3 (E) 5

9. Tom picked a natural number and multiplied it by 3. Which number CANNOT be the result of this multiplication?

(A) 987 (B) 444 (C) 204 (D) 105 (E) 103

10. How many hours is half of a third part of a quarter of 24 hours?

(A) $\frac{1}{3}$ (B) $\frac{1}{2}$ (C) 1 (D) 2 (E) 3

PROBLEMS 2005

Problems 4 points each

11. Eva cut a paper napkin into 10 pieces. She then also cut one of the pieces into 10 pieces. She repeated this process two more times. Into how many pieces did she cut the napkin?

   (A) 27   (B) 30   (C) 37   (D) 40   (E) 47

12. Mowgli usually walks from home to the beach and returns on an elephant. It takes him 40 minutes altogether. One day he traveled on the elephant from home to the beach and back which took him 32 minutes. How much time would he need to travel the same distance on foot both ways?

   (A) 24 minutes   (B) 42 minutes   (C) 46 minutes   (D) 48 minutes   (E) 50 minutes

13. A rectangular garden with an area of 30 m² was divided into three rectangular sections of flowers, vegetables, and strawberries (some of the dimensions are shown in the diagram). What is the area of the vegetable section if the flower section has an area of 10 m²?

   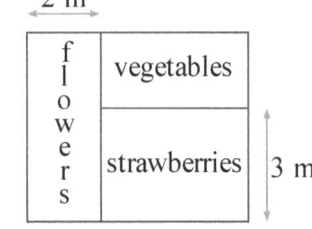

   (A) 4 m²   (B) 6 m²   (C) 8 m²   (D) 10 m²   (E) 12 m²

14. Grandpa suggested dividing all the peanuts between the family members in the following way: one person would get 5 kg, two people would get 4 kg each, four people would get 2 kg each, two people would get 1.5 kg each, and one person would not get any peanuts. Grandma suggested dividing the peanuts equally among all the family members. How many people would get more peanuts in the division suggested by Grandma than in the division suggested by Grandpa?

   (A) 3   (B) 4   (C) 5   (D) 6   (E) 7

15. How many two-digit numbers are there which can be expressed by using only different odd digits?

   (A) 15   (B) 20   (C) 25   (D) 30   (E) 50

16. Which of the cubes below is represented by the net of the cube shown to the right?

   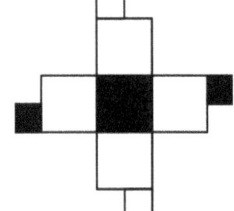

   (A)    (B) 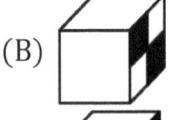   (C)

   (D)    (E)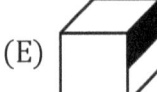

PROBLEMS 2005

17. The sum of five consecutive natural numbers is equal to 2005. The greatest of these numbers is:

(A) 401 (B) 403 (C) 404 (D) 405 (E) 2001

18. The number of all divisors of the number 100 is:

(A) 3 (B) 6 (C) 7 (D) 8 (E) 9

19. The frame of a rectangular painting was made out of wooden pieces of the same width. What is the width of those pieces if the outer perimeter of the frame is 8 decimeters longer than the inner perimeter?

(A) 4 decimeters (B) 6 decimeters (C) 1 decimeter (D) 8 decimeters
(E) The width depends on the dimensions of the painting.

20. How many more triangles than squares are shown in the picture?

(A) 4 more (B) 2 more (C) 1 more (D) 5 more (E) 3 more

Problems 5 points each

21. There are five containers in a treasure chest, in each container there are three boxes, and in each box there are 10 golden coins. The treasure chest, the containers, and the boxes are all locked. How many locks do you need to open to get 50 coins?

(A) 5 (B) 7 (C) 9 (D) 6 (E) 8

22. What number should replace x if we know that the number in any given circle above is the sum of the numbers in the two circles right below?

(A) 32 (B) 50 (C) 55 (D) 82 (E) 100

23. In a two-digit number, a is the tens digit and b is the ones digit. Which of the conditions below ensures that the number will be divisible by 6?

(A) $a + b = 6$ (B) $b = 6a$ (C) $b = 5a$ (D) $b = 2a$ (E) $a = 2b$

24. A wooden cube with the length of its side equal to 3 ft was painted with 0.25 gallon of paint. The cube was then cut up into unit cubes (with a side length of 1 ft). How much paint is needed to paint the unpainted sides of the unit cubes?

(A) 1.25 gallon (B) 1 gallon (C) 0.75 gallon (D) 0.5 gallon (E) 0.25 gallon

25. Five circles have radii of the same length. Four of them are touching the fifth circle and their centers are the vertices of a square (see the picture). The ratio of the area of the shaded region of the circles to the area of unshaded regions of the circles is:

(A) 1 : 3   (B) 1 : 4   (C) 2 : 5   (D) 2 : 3   (E) 5 : 4

26. From noon until midnight, Wise Cat sleeps under a chestnut tree. From midnight until noon he is awake telling stories. There is a note on that tree which says: "Two hours ago, Wise Cat was doing the same thing that he will be doing in an hour." For how many hours out of 24 hours is the note true?

(A) 6   (B) 12   (C) 18   (D) 3   (E) 21

27. Mark has 42 cubes with a side length of 1 cm. He used them to construct a prism, the base of which has a perimeter of 18 cm. The height of that prism is:

(A) 6 cm   (B) 5 cm   (C) 4 cm   (D) 3 cm   (E) 2 cm

28. Peter wrote on the board all three-digit numbers that have the following properties: the digits of each of the numbers are different, and the first is the square of the quotient of the second digit and the third digit. How many numbers did Peter write?

(A) 1   (B) 2   (C) 3   (D) 4   (E) 8

29. The equilateral triangle ABC (all sides congruent) has an area equal to 1. A bigger triangle was constructed out of 49 of these triangles (see the picture). The area of the shaded region is equal to:

(A) 20   (B) 22.5   (C) 23.5   (D) 25   (E) 32

30. Mary, Dorothy, Sylvia, Ella, and Kathy are sitting on a bench in the park. Mary is not sitting the farthest to the right; Dorothy is not sitting the farthest to the left. Sylvia is not sitting the farthest to the left nor the farthest to the right. Kathy is not sitting next to Sylvia, and Sylvia is not sitting next to Dorothy. Ella is sitting to the right of Dorothy, but not necessarily next to her. Which girl is sitting the farthest to the right?

(A) Kathy   (B) Dorothy   (C) Sylvia   (D) Ella   (E) It cannot be determined.

Problems from Year 2007

Problems 3 points each

1. Evaluate $2007 \div (2 + 0 + 0 + 7) - 2 \times 0 \times 0 \times 7$

    (A) 1 (B) 9 (C) 214 (D) 223 (E) 2007

2. Donald's grandfather sleeps through exactly a quarter of the day. Donald sleeps one and a half time as long as his grandfather. What fraction of the day does Donald spend sleeping?

    (A) $\frac{1}{6}$ (B) $\frac{1}{2}$ (C) $\frac{1}{8}$ (D) $\frac{3}{8}$ (E) $\frac{3}{4}$

3. Two angles of a certain triangle measure 12° and 48°. What is the measure of the third angle?

    (A) 30° (B) 300° (C) 100° (D) 120° (E) 90°

4. Each jump a certain kangaroo makes takes the same amount of time. If it takes the kangaroo 6 seconds to make 4 jumps, how long will it take the kangaroo to make 10 jumps?

    (A) 10 seconds (B) 12 seconds (C) 15 seconds (D) 18 seconds (E) 20 seconds

5. Adam's digital watch shows 16:07. He observes that the sum of the digits of the hour part and the sum of the digits of the minute part are equal. How many times will this be true between 16:00 and 17:00?

    (A) 10 (B) 6 (C) 7 (D) 8 (E) 3

6. One of the following five pieces forms a rectangle when placed next to the piece that is shown to the right. Which one is it?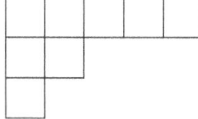

    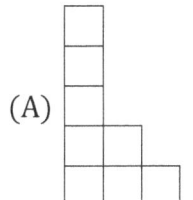 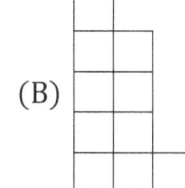 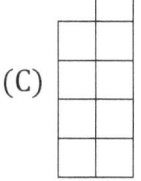 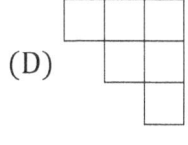 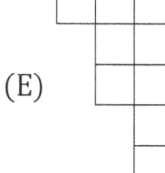
    (A)   (B)   (C)   (D)   (E)

7. The square shown in the picture must be filled in such a way that each of the digits 1, 2, and 3 appears in each row and in each column once and only once. If Harry started to fill in the square as shown, in how many ways he can complete the task?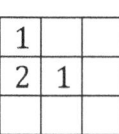

    (A) 1 (B) 2 (C) 3 (D) 4 (E) 5

PROBLEMS 2007

8. Ian was born on January 1st, 2002, and is older than Peter by 1 day less than 1 year. What is the date of Peter's birth?

   (A) January 2nd, 2003   (B) January 2nd, 2001   (C) December 31st, 2000
   (D) December 31st, 2002   (E) December 31st, 2003

9. Alexa is 8 years older than her two sisters, who are twins. The sum of the ages of all three girls is 32 years. How old is Alexa?

   (A) 12   (B) 18   (C) 16   (D) 20   (E) 14

10. What will be the height of a structure obtained by cutting a 1 meter × 1 meter × 1 meter cube into cubes having an edge length of 1 decimeter and then stacking all the decimeter cubes one on top of another? (1 m = 1 meter = 100 centimeters and 1 decimeter = 10 centimeters)

    (A) 100 m   (B) 1 km   (C) 10 km   (D) 1000 km   (E) 10 m

Problems 4 points each

11. Wanda cut a square sheet of paper with a perimeter of 20 inches into two rectangles. One of the rectangles has a perimeter of 16 inches. What is the perimeter of the other rectangle?

    (A) 8 inches   (B) 9 inches   (C) 12 inches   (D) 14 inches   (E) 16 inches

12. A square grid is composed of smaller squares. The two diagonals of this larger square grid were drawn, and then Hanna colored all the smaller squares which the diagonals passed through. How many squares make up the larger grid if she colored a total of 9 squares?

    (A) 9   (B) 16   (C) 25   (D) 64   (E) 81

13. Alex, Ben, Carl, and Daniel each participates in a different sport: karate, soccer, volleyball, and judo. Alex does not like sports played with a ball. Ben practices judo and often attends soccer games to watch his friend play. Which of the following statements could be true?

    (A) Alex plays volleyball.   (B) Ben plays soccer.   (C) Carl plays volleyball.
    (D) Daniel does karate.   (E) Alex does judo.

14. By what amount will the surface area of the rectangular block, shown in the picture, decrease if a rectangular section is removed as shown?

    (A) 27   (B) 54   (C) 72   (D) 108   (E) 126

15. Kate has a rectangular strip of cardboard 27 inches long which she divides into 4 sections of different lengths. She then draws two line segments connecting the centers of two adjacent rectangles as shown in the diagram. Find the sum of the lengths of these two line segments.

(A) 12 inches (B) 13.5 inches (C) 14 inches (D) 14.5 inches
(E) The sum depends on the lengths of the four sections.

16. 60 birds were sitting in 3 trees. Suddenly, some of the birds flew away, and the same number of birds was left on each tree. If 6 birds flew away from the first tree, 8 birds flew away from the second tree, and 4 birds flew away from the third tree, how many birds were there on the second tree at the beginning?

(A) 26 (B) 24 (C) 22 (D) 21 (E) 20

17. Two squares, each measuring 9 × 9 inches, are placed overlapping each other so that a 9 × 13 inch rectangle is created. Find the area of the region where the two squares overlap.

(A) 36 in² (B) 45 in² (C) 54 in² (D) 63 in² (E) 72 in²

18. Ian released a homing pigeon at 7:30 a.m. The pigeon arrived at its destination at 9:10 a.m. How many miles did the pigeon travel if it flies 4 miles in 10 minutes?

(A) 14 (B) 20 (C) 40 (D) 56 (E) 64

19. A mechanical kangaroo moves on the board shown in the picture. It begins in square A2 and goes in the forward direction indicated by the arrow. The kangaroo cannot leave the board or enter the shaded squares. It is programmed to move forward one square at a time until it meets an obstacle, in which case it makes a 90° right turn and continues in the new forward direction. It will stop if it cannot continue moving forward after making a right turn. On which square will the kangaroo stop?

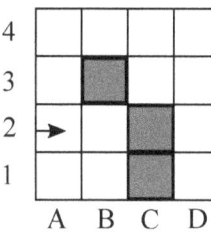

(A) B2 (B) A1 (C) C3 (D) D1 (E) It will never stop.

20. Anna is 10 years old and her mother is 4 times as old as Anna. How old will Anna's mother be when Anna is twice as old as she is right now?

(A) 40 (B) 50 (C) 60 (D) 70 (E) 80

PROBLEMS 2007

Problems 5 points each

21. Diagonals are drawn on three adjacent faces of a cube as shown in the picture. From which of the grids below could such a cube be constructed?

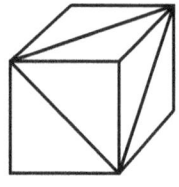

(A) 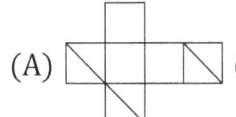 (B) 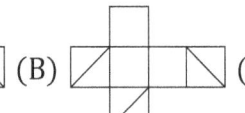 (C)  (D)  (E)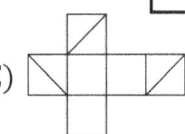

22. To the right of a certain two-digit number the same number has been written, creating a four-digit number. How many times is the new four-digit number greater than the original two-digit number?

    (A) 100   (B) 101   (C) 1000   (D) 1001   (E) 10

23. A parallelogram is divided into two parts, P_1 and P_2, as shown. Which of the following statements is true?

    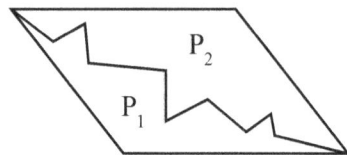

    (A) P_2 has a larger perimeter than P_1.
    (B) P_2 has a smaller perimeter than P_1.
    (C) P_2 has a smaller area than P_1.
    (D) P_2 and P_1 have equal perimeters.
    (E) P_2 and P_1 have equal areas.

24. What is the 2007th letter in the sequence *KANGAROOKANGAROOKANG*...?

    (A) *K*   (B) *A*   (C) *N*   (D) *R*   (E) *O*

25. Figure A is made up of 4 rectangles, each 10 inches wide. Each rectangle is 25 inches longer than the previous one. Figure B is made by rearranging the rectangles that make up figure A. By how much is the perimeter of B greater than the perimeter of A?

    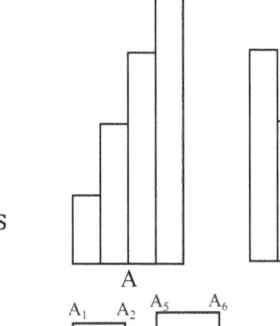

    (A) 10 inches   (B) 20 inches   (C) 25 inches   (D) 40 inches
    (E) 50 inches

26. In the figure shown to the right, squares are formed by dividing the line segment AB, which is 24 inches long. What is the length of the thin line $AA_1A_2A_3 \ldots A_{10}A_{11}A_{12}B$?

    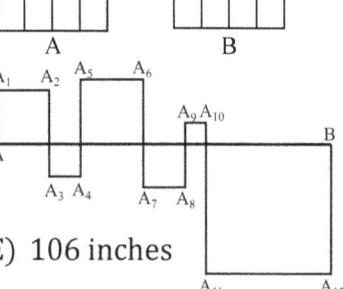

    (A) 48 inches   (B) 72 inches   (C) 96 inches   (D) 56 inches   (E) 106 inches

PROBLEMS 2007

27. Tom thought of an integer, which Robert then multiplied by either 5 or 6. Then Dan added either 5 or 6 to Robert's new number and Adam subtracted either 5 or 6 from Dan's new number to get 73. What integer did Tom think of?

   (A) 10   (B) 11   (C) 12   (D) 14   (E) 15

28. The square to the right has a side 10 inches long. Angle *EAB* measures 75 degrees and angle *ABE* measures 30 degrees. What is the length of segment *EC*?

   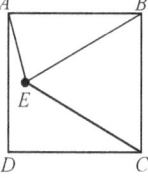

   (A) 8 inches   (B) 9 inches   (C) 9.5 inches   (D) 10 inches   (E) 11 inches

29. In the figure, squares *ABCD* and *EFGH* are such that: $AB = EF$ and $\overline{AB} \parallel \overline{EF}$. The area of the shaded region is 1. What is the area of *ABCD*?

   (A) 1   (B) 2   (C) $\frac{1}{2}$ (D) $\frac{3}{2}$
   (E) It cannot be determined.

30. On a die, the sum of the dots on opposite faces is always 7. Four such identical dice make up the figure in the picture. The dice are arranged in such a way that the faces which touch each other have the same number of dots. How many dots are on the face marked with the question mark?

   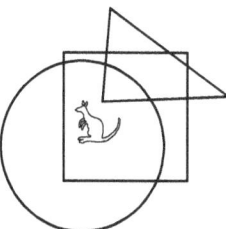

   (A) 5   (B) 6   (C) 2   (D) 3
   (E) Not enough information is given.

Problems from Year 2009

Problems 3 points each

1. Which of the numbers below is even?

   (A) 2009   (B) $2 + 0 + 0 + 9$ (C) $200 - 9$ (D) 200×9 (E) $200 + 9$

2. Where is the kangaroo?

   (A) In the circle and in the triangle, but not in the square.
   (B) In the circle and in the square, but not in the triangle.
   (C) In the triangle and in the square, but not in the circle.
   (D) In the circle, but not in the square and not in the triangle.
   (E) In the square, but not in the circle and not in the triangle.

3. How many integers are there between 2.009 and 19.03?

    (A) 16 (B) 17 (C) 14 (D) 15 (E) more than 17

4. The smallest number of digits that need to be erased from the number 12323314 in order to get a number that reads the same from left to right as from right to left is

    (A) 1 (B) 2 (C) 3 (D) 4 (E) 5

5. There are three boxes on the table: one white, one red, and one green. One of them contains only a chocolate bar, another contains only an apple, and the third one is empty. What is the color of the box which contains the chocolate bar, if we know that the chocolate bar is either in the white or in the red box, and the apple is neither in the white nor in the green box?

    (A) white (B) red (C) green (D) none of these
    (E) It is impossible to determine.

6. The picture shows square *KLMN* and equilateral triangle *KLP*. The point where the diagonal *KM* of the square and the side *LP* of the triangle intersect has been marked with *Q*. What is the measure of ∠*LQM*?

    (A) 95° (B) 105° (C) 115° (D) 125° (E) 135°

7. A bridge is built across a river. The river is 120 meters wide. One quarter of the bridge is over land on the left bank of the river and one quarter of the bridge is over land on the right bank of the river. How long is the bridge?

    (A) 150 meters (B) 180 meters (C) 210 meters
    (D) 240 meters (E) 270 meters

8. A rectangle is built out of squares of three different sizes as shown in the picture. The side of the smallest square is 20 cm long. What is the length of the line marked in bold?

    (A) 380 cm (B) 400 cm (C) 420 cm (D) 440 cm (E) 1680 cm

9. There are cats and dogs in a room. The number of cat paws in this room is twice the number of dog noses. The number of cats is

    (A) twice the number of dogs
    (B) equal to the number of dogs
    (C) half the number of dogs
    (D) ¼ the number of dogs
    (E) ⅙ the number of dogs

PROBLEMS 2009

10. Using identical sticks, we can form digits as shown in the picture. Following this example, John was making all of the two-digit numbers by using such sticks. Stan was writing down the number of sticks used to form each two-digit number. The greatest number written by Stan is:

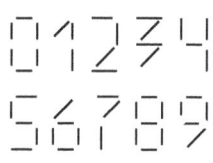

(A) 99 (B) 11 (C) 12 (D) 13 (E) 14

Problems 4 points each

11. How many positive whole numbers n have the property that $n + 2$ is a divisor of 78?

(A) 7 (B) 8 (C) 6 (D) 5 (E) 9

12. The quadrilateral $ABCD$ has sides $|AB| = 11$, $|BC| = 7$, $|CD| = 9$, and $|DA| = 3$, with right angles at A and C. What is the area of this quadrilateral?

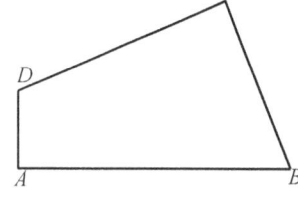

(A) 30 (B) 44 (C) 48 (D) 52 (E) 60

13. A certain dance group started out with 39 boys and 23 girls. Every week, 6 more boys and 8 more girls joined the dance group. After a few weeks, the number of boys and the number of girls in the dance group was equal. How many participants were in the dance group at that time?

(A) 144 (B) 154 (C) 164 (D) 174 (E) 184

14. Two rectangles with dimensions 8 × 10 and 9 × 12 partly overlap, as shown in the picture. The area of the shaded region is 37. What is the area of the region covered with diagonal lines?

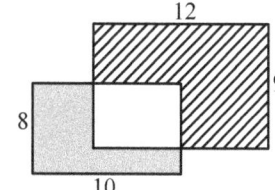

(A) 60 (B) 62 (C) 62.5 (D) 64 (E) 65

15. The numbers 1 to 8 are written on eight cards, a different number on each card. The cards are then divided into groups A and B in such a way that the sum of the numbers on the cards in group A is equal to the sum of the numbers on the cards in group B. If we know that group A contains only three cards, then we know for sure that:

(A) exactly three cards in group B have odd numbers on them.
(B) four of the cards in group B have even numbers on them.
(C) the card with 1 is not in group B.
(D) the card with 2 is in group B.
(E) the card with 5 is in group B.

PROBLEMS 2009

16. The equilateral triangle, the rectangle, and the square that make up the figure shown in the picture all have perimeters which are equal. The length of one side of the square is 9 cm. What is the length of the shorter side of the rectangle?

   (A) 4 cm   (B) 5 cm   (C) 6 cm   (D) 7 cm   (E) 8 cm

17. What is the smallest number of identical cubes that are needed to make a rectangular prism with the dimensions of 40 × 40 × 60?

   (A) 6   (B) 12   (C) 96   (D) 12,000   (E) 96,000

18. Adam wants to read a 290-page book. He decides to read 4 pages each day except Sunday, and to read 25 pages each Sunday. He started reading the book on a Sunday. How many days will it take him to read it?

   (A) 15   (B) 46   (C) 40   (D) 35   (E) 41

19. Adam, Bart, Caesar, and Daniel took the first four places (1st, 2nd, 3rd, and 4th) in a chess tournament. The sum of the numbers of the places of Adam, Bart, and Daniel is equal to 6, and the sum of the numbers of places of Bart and Caesar is also equal to 6. It is also known that Bart did better than Adam. Which of the boys won the first place?

   (A) Adam   (B) Bart   (C) Caesar   (D) Daniel   (E) It cannot be determined.

20. How many different rectangles are there that have an area of 2009 and sides with lengths that are whole numbers? (If two rectangles can be placed on top of each other in such a way that they overlap completely, they are not considered different.)

   (A) 1   (B) 2   (C) 3   (D) 5   (E) 10

Problems 5 points each

21. Jane multiplied the product of 18 factors, each equal to 8, by the product of 50 factors, each equal to 5. How many digits does her final product have?

   (A) 13   (B) 40   (C) 52   (D) 60   (E) 100

22. Of the four statements given below about a natural number n, two are true and two are false.
    - n is divisible by 5
    - n is divisible by 11
    - n is divisible by 55
    - n is less than 10

    The number n is

    (A) 0 (B) 5 (C) 10 (D) 11×55 (E) 55

23. The picture shows a solid formed with 6 triangular faces. At each vertex there is a number, and the sum of the three numbers at the vertices of each face is the same. Two of the numbers, 1 and 5, are shown in the picture. What is the sum of all the numbers at the vertices of the solid?

    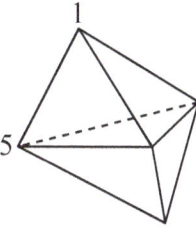

    (A) 9 (B) 12 (C) 17 (D) 18 (E) 24

24. A certain hotel has 5 floors. There are 35 rooms on each floor. Each of the rooms is numbered with a three-digit number. The first digit indicates the floor number, and the remaining two digits form the room number on the given floor. The rooms on each floor are numbered in order. For example, all the rooms on the third floor are numbered from 301 to 335. How many times was the digit 2 used in numbering all the rooms in this hotel?

    (A) 60 (B) 65 (C) 95 (D) 100 (E) 105

25. $ABCD$ is a square with side lengths equal to 10 cm. The distance from point N to point M is 6 cm. Each of the non-shaded regions represents one of four identical isosceles right triangles or one of four identical squares. What is the area of the shaded region inside the square $ABCD$?

    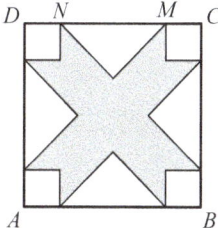

    (A) 42 cm² (B) 46 cm² (C) 48 cm² (D) 52 cm² (E) 58 cm²

26. The picture shows a table with symbols a, b, and c, each of which represents a number. The sums of the numbers in each row and in each column are given. What is the value of the expression $a + b - c$?

    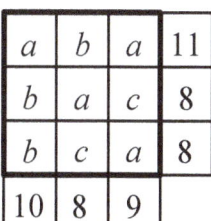

    (A) 4 (B) 5 (C) 6 (D) 7 (E) 8

27. A complete set of dominoes contains 28 pieces. The pieces show every possible combination of two numbers of dots from 0 to 6 inclusive. How many dots are there altogether in a complete set of dominoes?

    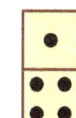

    (A) 84 (B) 105 (C) 126 (D) 147 (E) 168

28. In a 4 × 2 table, two numbers are written in the first row. Each of the rows below contains the sum and the difference of the numbers written in the previous row (see the picture for an example). In a 7 × 2 table, filled in the same way, the numbers in the last row are 96 and 64. What is the sum of the numbers in the first row of this table?

    (A) 8   (B) 10   (C) 12   (D) 20   (E) 24

29. In the land of Funnyfeet, the left foot of each person is either one or two sizes bigger than his or her right foot. However, shoes are always sold in pairs of the same size, and only in the whole sizes. A group of friends decided to buy shoes, and to save money they bought shoes together. After they all put on the shoes that fit them, there were exactly two shoes left over, one of size 36 and another of size 45. What is the smallest possible number of people in this group of friends?

    (A) 5   (B) 6   (C) 4   (D) 9   (E) 8

30. We want to color the squares in the grid using colors A, B, C, and D in such a way that neighboring squares do not have the same color (squares that share a side or a vertex are considered neighboring squares). Four of the squares have been colored as shown. How can the shaded square be colored?

    (A) color A only   (B) color B only   (C) color C only   (D) color D only
    (E) There are two different possibilities.

Problems from Year 2011

Problems 3 points each

1. Basil is writing the word KANGAROO, one letter each day. He wrote the first letter on Wednesday. On what day will he finish writing the word?

    (A) Monday   (B) Tuesday   (C) Wednesday   (D) Thursday   (E) Friday

2. A motorcyclist drove a distance of 28 kilometers in 30 minutes. How many kilometers would he drive in one hour if he drives at the same speed?

    (A) 28 km   (B) 56 km   (C) 36 km   (D) 58 km   (E) 62 km

3. How many faces do six cubes have altogether?

    (A) 18   (B) 16   (C) 24   (D) 36   (E) 48

PROBLEMS 2011

4. A square piece of paper was cut into two pieces along a straight line. Which of the following shapes cannot be the result of the cut?

   (A) square (B) rectangle (C) right triangle
   (D) pentagon (E) isosceles triangle

5. In Crazytown, the houses on the right side of Number Street have odd numbers. However, Crazytowners do not use numbers containing the digit 3. The first house on the right side of the street is numbered 1. What is the number of the fifteenth house on the right side of the street?

   (A) 29   (B) 41   (C) 43   (D) 45   (E) 47

6. A system of pipes connects the upper container Z with two lower containers X and Y (see the picture). At each fork in the pipes, the water flowing down divides into two equal amounts. 1000 gallons of water were poured into the empty container Z. How many gallons of water will reach container Y?

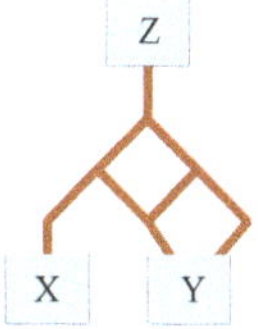

   (A) 800   (B) 750   (C) 666.67   (D) 660   (E) 500

7. Square $ABCD$ has a side equal to 5 in. Point P is 5 in away from point A and 1 in away from side BC. What is the area of triangle APD?

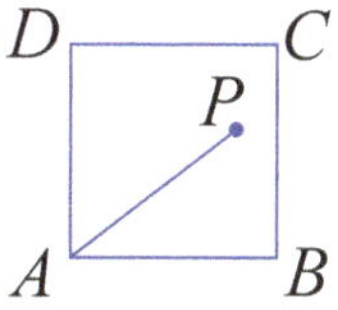

   (A) 8 in^2 (B) 10 in^2 (C) 25 in^2 (D) 16 in^2 (E) 15 in^2

8. In triangle ABC, $|AC| = |BC|$ and $\angle ACB = 30°$. AD is a height of the triangle, as marked in the picture. What is the measure of angle BAD?

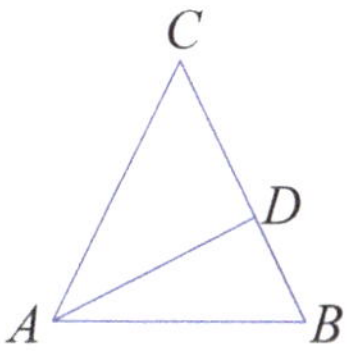

   (A) 30°   (B) 25°   (C) 20°   (D) 15°   (E) 10°

9. Which of the pieces below is needed to make the solid shown to the right into a prism?

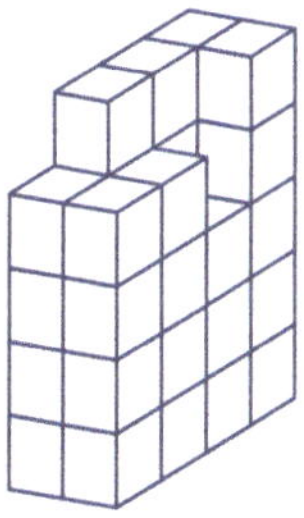

   (A) 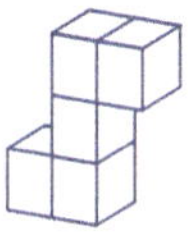   (B) 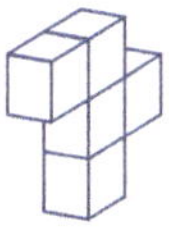   (C) 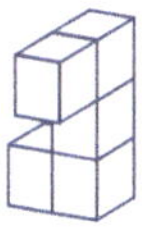   (D) 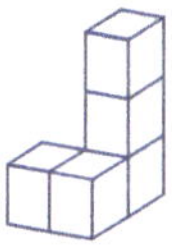   (E)

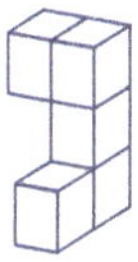

10. Maria's cat drinks 60 ml of milk each day. However, if the cat catches a mouse, he drinks one third more milk. During the last two weeks, the cat caught a mouse each day. How much milk did the cat drink in the last two weeks?

    (A) 840 ml   (B) 980 ml   (C) 1050 ml   (D) 1120 ml   (E) 1960 ml

© Math Kangaroo in USA, NFP   www.mathkangaroo.org

Problems 4 points each

11. Andrew wrote the letters of the word KANGAROO in a table with 2 rows and 4 columns, with one letter in each cell. He can write the first letter in any cell. He then writes each of the following letters into a cell that has at least one point in common with the cell in which the letter before it was written. Which of the tables shown below cannot have been filled in by Andrew?

    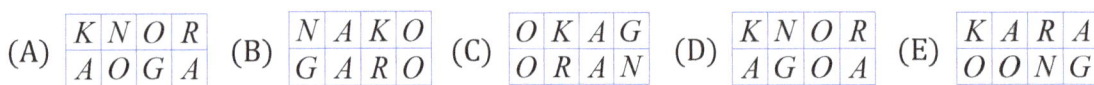

    (A) $\begin{array}{|c|c|c|c|} \hline K & N & O & R \\ \hline A & O & G & A \\ \hline \end{array}$
    (B) $\begin{array}{|c|c|c|c|} \hline N & A & K & O \\ \hline G & A & R & O \\ \hline \end{array}$
    (C) $\begin{array}{|c|c|c|c|} \hline O & K & A & G \\ \hline O & R & A & N \\ \hline \end{array}$
    (D) $\begin{array}{|c|c|c|c|} \hline K & N & O & R \\ \hline A & G & O & A \\ \hline \end{array}$
    (E) $\begin{array}{|c|c|c|c|} \hline K & A & R & A \\ \hline O & O & N & G \\ \hline \end{array}$

12. 2011 and all four-digit numbers that can be made by rearranging the digits in 2011 were placed in increasing order. What is the difference between the two numbers on either side of 2011?

    (A) 890   (B) 891   (C) 900   (D) 909   (E) 990

13. Four of the numbers in the box on the left were moved to spaces in the box on the right in such a way as to make the addition correct (see the picture). Which number remains in the box on the left?

    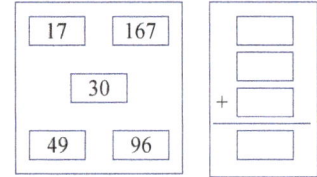

    (A) 17   (B) 30   (C) 49   (D) 96   (E) 167

14. Agatha used 36 identical cubes to build a fence around a square region. The cubes are all connected. How many cubes identical to the cubes she used to build the fence does Agatha need to fill in the enclosed region?

    (A) 36   (B) 49   (C) 64   (D) 81   (E) 100

15. Paul wanted to multiply a certain number by 301, but instead he multiplied it by 31. He got 372 as the result. What would the result have been if he had multiplied by the number he wanted to?

    (A) 3,010   (B) 3,612   (C) 3,702   (D) 3,720   (E) 30,720

16. There are four identical right triangles inside a rectangle, as shown in the picture. The lengths of the two sides of the rectangle are 28 cm and 30 cm. What is the sum of the areas of all four triangles?

    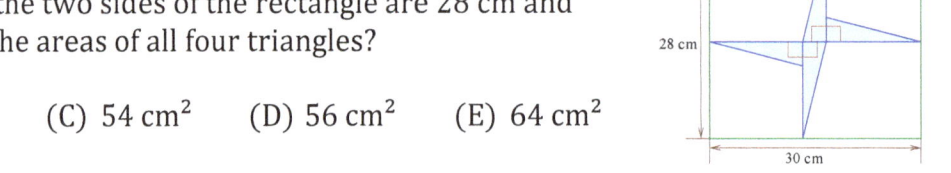

    (A) 46 cm²   (B) 52 cm²   (C) 54 cm²   (D) 56 cm²   (E) 64 cm²

PROBLEMS 2011

17. In a tournament, the soccer team FC Barcelona scored a total of three goals and had one goal scored against it. It is known that this team won one game, tied one game, and lost one game. What was the score of the game which FC Barcelona won?

(A) 2 : 0   (B) 3 : 0   (C) 1 : 0   (D) 4 : 1   (E) 0 : 1

18. There are three points that form a triangle. In how many ways can a fourth point be chosen in order to create a parallelogram?

(A) 1   (B) 2   (C) 3   (D) 4   (E) 5

19. 8 points are connected with several segments as shown in the picture. Each of these points needs to be labeled with number 1, 2, 3, or 4. The numbers on each end of any given segment need to be different. Three of the points are already labeled (see the picture). How many points will be labeled with 4?

(A) 1   (B) 2   (C) 3   (D) 4   (E) 5

20. There are 10 students in a dance class. One of the boys brought 80 pieces of candy. If he gives each of the girls in his class the same number of pieces of candy, there will be 3 pieces of candy left. How many boys are there in the class if we know that there are at least 2 girls?

(A) 1   (B) 2   (C) 3   (D) 5   (E) 7

Problems 5 points each

21. Eve wants to make a square using only pieces like the one in the picture. What is the smallest number of pieces she needs to make a square?

(A) 8   (B) 10   (C) 12   (D) 16   (E) 20

22. By arranging four puzzle pieces: different shapes can be made. Which of the five shapes below cannot be made out of these four pieces?

(A)   (B)   (C)   (D)   (E)

23. In how many different ways can we choose four of the numbers 2, 3, 5, 6, 10, 15, and 30 in such a way that any two out of the four chosen numbers will have a common factor greater than 1?

(A) 1   (B) 3   (C) 4   (D) 6   (E) 7

24. We have several square grids with odd numbers of rows and columns. All the small squares in the grid which are either in a row or in a column with an even number are painted white. The rest of the small squares are painted black. Grids that are 1 × 1, 3 × 3, and 5 × 5 are shown in the picture. How many small white squares will there be in a grid that has 25 small black squares?

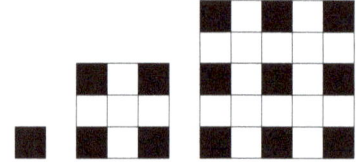

(A) 25   (B) 39   (C) 45   (D) 56   (E) 72

25. The rectangle $ABCD$ is divided into 9 squares. The areas of two of the squares are 64 in² and 81 in², as shown in the picture. What is the length of side AB ?

(A) 32 in   (B) 33 in   (C) 38 in   (D) 39 in   (E) 36 in

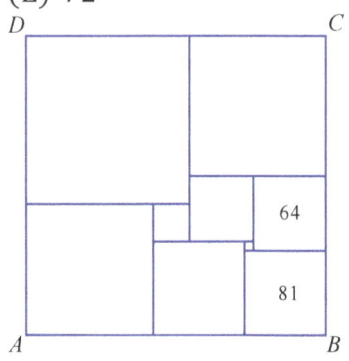

26. In a certain month there were 5 Saturdays and 5 Sundays, but only 4 Fridays and 4 Mondays. In the following month there will be

(A) 5 Wednesdays.   (B) 5 Thursdays.   (C) 5 Fridays.
(D) 5 Saturdays.   (E) 5 Sundays.

27. On a six-sided die, the number of dots on any two opposite faces adds up to 7. The picture shows three six-sided dice stacked one on top of the other in such a way that the sum of the number of dots on any two faces that touch is 5. In the picture, the number of dots on one of the faces is shown. How many dots are there on the top face of the solid?

(A) 2   (B) 3   (C) 4   (D) 5   (E) 6

28. Adam is drawing sets of 4 circles in such a way that in each set any 2 circles have exactly one point in common. In each set, he counts the number of points which lie on at least two circles. The greatest number of such points in a set is

(A) 1   (B) 4   (C) 5   (D) 6   (E) 8

29. Alex said that Thomas was lying. Thomas said that Mark was lying. Mark said that Thomas was lying. Tony said that Alex was lying. How many of the boys were lying?

(A) 0   (B) 1   (C) 2   (D) 3   (E) 4

30. How many five-digit numbers which use the digits 1, 2, 3, 4, and 5, with all different digits, have the following properties when looking at each number from left to right: the first two digits form a number divisible by 2, the first three digits form a number divisible by 3, the first four digits form a number divisible by 4, and the five-digit number is divisible by 5?

(A) There are no such numbers. (B) 1 (C) 2 (D) 5 (E) 10

Problems from Year 2013

Problems 3 points each

1. We put 2, 0, 1, 3 into an adding machine, as shown. What is the result in the box with the question mark?

   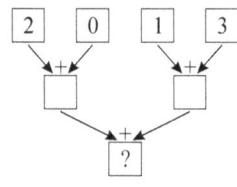

   (A) 2 (B) 3 (C) 4 (D) 5 (E) 6

2. Nathalie wanted to build the same cube as Diana had (Figure 1). However, Nathalie ran out of small cubes and built only a part of the cube, as you can see in Figure 2. How many small cubes must be added to Figure 2 to form Figure 1?

   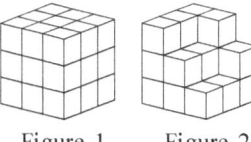

   Figure 1   Figure 2

   (A) 5 (B) 6 (C) 7 (D) 8 (E) 9

3. Find the distance which Mara covers to get to her friend Bunica.

   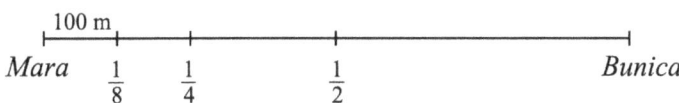

   (A) 300 m (B) 400 m (C) 800 m (D) 1 km (E) 700 m

4. Nick is learning to drive. He knows how to turn right but cannot turn left. What is the smallest number of turns he must make in order to get from A to B, starting in the direction of the arrow?

   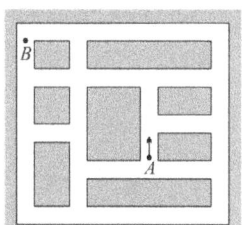

   (A) 3 (B) 4 (C) 6 (D) 8 (E) 10

5. The sum of the ages of Ann, Bob, and Chris is 31 years. What will the sum of their ages be in three years?

   (A) 32 (B) 34 (C) 35 (D) 37 (E) 40

PROBLEMS 2013

6. What digit must be placed in all three boxes of ☐☐ × ☐ = 176 in order to make the multiplication work?

   (A) 6   (B) 4   (C) 7   (D) 9   (E) 8

7. Michael has to take a pill every 15 minutes. He took the first pill at 11:05. What time did he take the fourth pill?

   (A) 11:40   (B) 11:50   (C) 11:55   (D) 12:00   (E) 12:05

8. By drawing two circles, Mike obtained a figure which consists of three regions (see picture). At most, how many regions could he obtain by drawing two squares?

   (A) 3   (B) 5   (C) 6   (D) 8   (E) 9

9. The number 36 has the property that it is divisible by the digit in the ones position, because 36 is divisible by 6. The number 38 does not have this property. How many numbers between 20 and 30 have this property?

   (A) 2   (B) 3   (C) 4   (D) 5   (E) 6

10. Ann has a lot of pieces like the one shown on the top of the picture. She tries to put as many as possible in the 4 by 5 rectangle, also shown. The pieces may not overlap each other. What is the largest possible number of pieces Ann can put in the rectangle?

    (A) 2   (B) 3   (C) 4   (D) 5   (E) 6

Problems 4 points each

11. Which of the following pieces covers the largest number of dots in the table?

    (A)   (B)   (C)   (D)   (E)

12. Mary shades various shapes on square sheets of paper, as shown.

    How many of these shapes have the same perimeter as the sheet of paper itself?

    (A) 2   (B) 3   (C) 4   (D) 5   (E) 6

PROBLEMS 2013

13. Ann rides her bicycle throughout the afternoon at a constant speed. She sees her watch at the beginning and at the end of the ride with the following result:

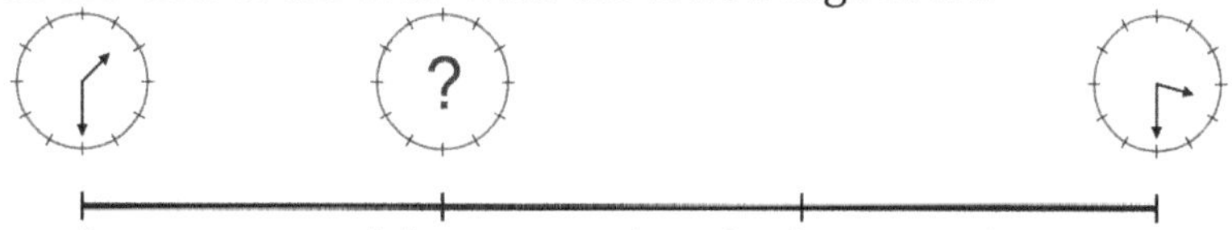

Which picture shows the position of the minute hand when Ann finishes one third of the ride?

(A)    (B)    (C)    (D) 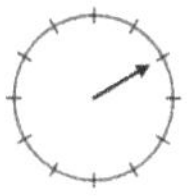   (E)

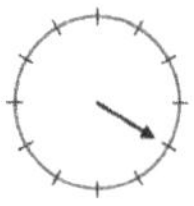

14. Matthew is catching fish. If he had caught three times as many fish as he actually did, he would have 12 more fish than he does. How many fish did he catch?

(A) 7   (B) 6   (C) 5   (D) 4   (E) 3

15. John made a building out of cubes. In the picture you see this building from above. In each cell you see the number of cubes in that particular tower. When you look from the front, what do you see?

BACK
4	2	3	2
3	3	1	2
2	1	3	1
1	2	1	2
FRONT

(A) 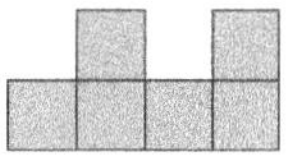   (B) 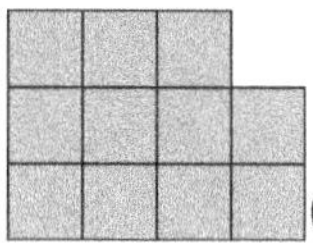   (C) 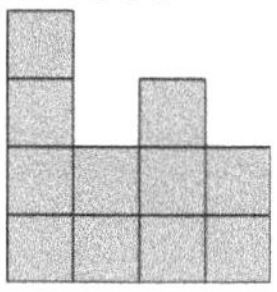   (D) 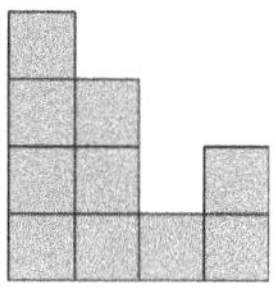   (E)

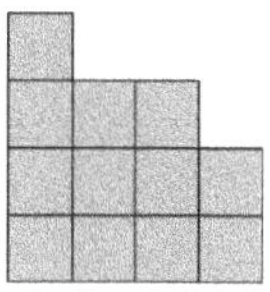

16. In an election each of the five candidates got a different number of votes. The candidates received 36 votes in total. The winner got 12 votes. The candidate in the last place got 4 votes. How many votes did the candidate in the second place get?

(A) 8   (B) 8 or 9   (C) 9   (D) 9 or 10   (E) 10

17. From a wooden cube with a side of 3 cm we cut out at the corner a little cube with a side of 1 cm (see picture). What is the number of faces of the solid after cutting out such a small cube at each corner of the big cube?

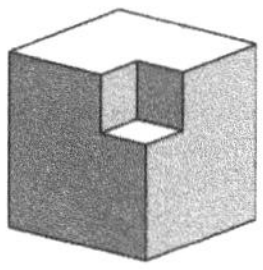

(A) 16   (B) 20   (C) 24   (D) 30   (E) 36

18. Find the number of pairs of two-digit natural numbers whose difference is equal to 50.

(A) 40   (B) 30   (C) 50   (D) 60   (E) 10

19. The final game of the field hockey championship was filled with goals. There were 6 goals in the first half and the guest team was leading after the first half. After the home team scored 3 goals in the second half, the home team won the game. How many goals did the home team score altogether?

(A) 3   (B) 4   (C) 5   (D) 6   (E) 7

© Math Kangaroo in USA, NFP   www.mathkangaroo.org

20. In the squares of the 4 × 4 board, numbers are written such that the numbers in adjacent squares differ by 1. Numbers 3 and 9 appear in the table. The number 3 is in the top left corner as shown. How many different numbers appear in the table?

(A) 4   (B) 5   (C) 6   (D) 7   (E) 8

Problems 5 points each

21. Aron, Bern, and Carl always lie. Each of them owns one stone, either a red stone or a green stone. Aron says: "My stone is the same color as Bern's stone." Bern says: "My stone is the same color as Carl's stone." Carl says: "Exactly two of us own red stones."
Which of the following statements is true?

(A) Aron's stone is green.
(B) Bern's stone is green.
(C) Carl's stone is red.
(D) Aron's stone and Carl's stone have different colors.
(E) None of the above is true.

22. 66 cats signed up for the contest MISS CAT 2013. After the first round, 21 were eliminated because they failed to catch a mouse. 27 cats out of those that remained in the contest had stripes and 32 of them had one black ear. All the striped cats with one black ear got to the final. What is the minimum number of finalists?

(A) 5   (B) 7   (C) 13   (D) 14   (E) 27

23. There are four buttons in a row as shown below. Two of them show happy faces, and two of them show sad faces. If we press on a face, its expression turns to the opposite (e.g., a happy face turns into a sad face). In addition to this, the adjacent buttons also change their expressions to the opposite. What is the least number of times you need to press the buttons in order to get all happy faces?

(A) 2   (B) 3   (C) 4   (D) 5   (E) 6

24. 40 boys and 28 girls stand in a circle, hand in hand, all facing inwards. Exactly 18 boys give their right hand to a girl. How many boys give their left hand to a girl?

(A) 18   (B) 9   (C) 28   (D) 14   (E) 20

25. A 2 × 2 × 2 cube is to be constructed using 4 white and 4 black unit cubes. How many different cubes can be constructed in this way? (Two cubes are not different if one can be obtained by rotating the other.)

(A) 16   (B) 9   (C) 8   (D) 7   (E) 6

26. How many 3-digit numbers possess the following property: after subtracting 297 from such a number, we get a 3-digit number consisting of the same digits in the reverse order?

(A) 6   (B) 7   (C) 10   (D) 60   (E) 70

27. When Matthew and Marten found their old model railway, Matthew quickly made a perfect circle from 8 identical track parts. Marten starts to make another track with two of these pieces as shown in the picture. He wants to use as few pieces as possible to make a closed track. How many pieces does his track consist of?

(A) 11   (B) 12   (C) 14   (D) 15   (E) 16

28. There were 2013 inhabitants on an island. Some of them were knights and the others were liars. The knights always tell the truth and the liars always lie. Every day, one of the inhabitants said: "After my departure the number of knights on the island will equal the number of liars," and then left the island. After 2013 days there was nobody on the island. How many liars were there initially?

(A) 0   (B) 1006   (C) 1007   (D) 2013   (E) It is impossible to determine.

29. Starting with a list of three numbers, the "change-sum" procedure creates a new list by replacing each number by the sum of the other two. For example, from {3, 4, 6} "change-sum" gives {10, 9, 7} and a new "change-sum" leads to {16, 17, 19}. If we begin with the list {20, 1, 3}, what is the maximum difference between two numbers of the list after 2013 consecutive "change-sums"?

(A) 1   (B) 2   (C) 17   (D) 19   (E) 2013

30. Alice forms 4 identical numbered cubes using the net shown. She then glues them together to form a 2 × 2 × 1 block as shown. Only faces with identical numbers are glued together. Alice then finds the total of all the numbers on the outside surface of the block. What is the largest total that Alice can get?

(A) 66   (B) 68   (C) 72   (D) 74   (E) 76

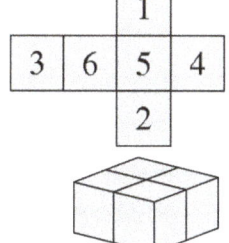

Problems from Year 2015

Problems 3 points each

1. Which figure has one half of its area shaded?

   (A) 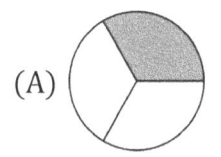 (B)  (C) 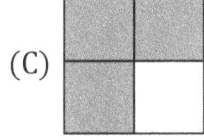 (D) 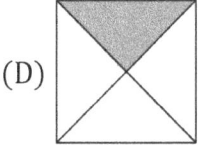 (E)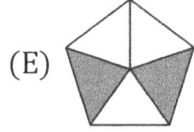

2. My umbrella has KANGAROO written on top. It is shown in the picture to the right. Which of the pictures below does not show my umbrella?

   (A)  (B)  (C)  (D)  (E)

3. Sam painted the 9 squares as shown in the figure to the right using the colors black, white, and gray. At least how many squares does he need to repaint so that no two squares with a common side are the same color?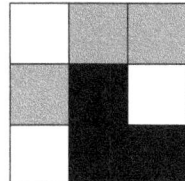

   (A) 2   (B) 3   (C) 4   (D) 5   (E) 6

4. There are 10 ducks. 5 of these ducks lay an egg every day. The other 5 lay an egg every other day. How many eggs do the 10 ducks lay in a period of 10 days?

   (A) 75   (B) 60   (C) 50   (D) 25   (E) 10

5. The figure to the right shows a board where each small square has an area of 4 cm². What is the length of the thick black line?

   (A) 16 cm   (B) 18 cm   (C) 20 cm   (D) 21 cm   (E) 23 cm

6. Which of the following improper fractions is smaller than 2?

   (A) $\frac{19}{8}$ (B) $\frac{20}{9}$ (C) $\frac{21}{10}$ (D) $\frac{22}{11}$ (E) $\frac{23}{12}$

7. A pumpkin and a watermelon together weigh 8 kg. The watermelon is 2 kg lighter than the pumpkin. How much does the pumpkin weigh?

   (A) 2 kg   (B) 3 kg   (C) 4 kg   (D) 5 kg   (E) 6 kg

PROBLEMS 2015

8. Each plant in John's garden has either 5 leaves only, or 2 leaves and 1 flower. In total, the plants have 6 flowers and 32 leaves. How many plants are there?

   (A) 10 (B) 12 (C) 13 (D) 15 (E) 16

9. Alva has 4 paper strips of the same length. She glues 2 of them together with a 10 cm overlap and gets a strip 50 cm long.

   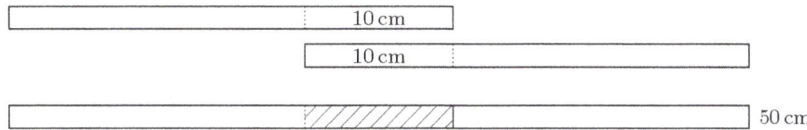

   With the other two paper strips, she wants to make a strip 56 cm long. How long should the overlap be?

   (A) 4 cm (B) 6 cm (C) 8 cm (D) 10 cm (E) 12 cm

10. Tom used 6 squares with a side length of 1 to form the shape shown in the picture. What is the perimeter of the shape?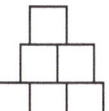

    (A) 9 (B) 10 (C) 11 (D) 12 (E) 13

Problems 4 points each

11. Every day Mary writes down the date and calculates the sum of the digits written. For example, on March 19 she writes 03/19 and calculates $0 + 3 + 1 + 9 = 13$. What is the largest sum that she calculates during a year?

    (A) 7 (B) 13 (C) 14 (D) 16 (E) 20

12. The rectangle $ABCD$ in the picture consists of 4 equal rectangles. If BC has a length of 1 cm, what is the length of AB?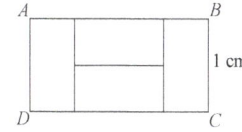

    (A) 4 cm (B) 3 cm (C) 2 cm (D) 1 cm (E) 0.5 cm

13. Which of these five nets cannot be the net of a pyramid? (A net is a flat shape that can be folded up to form a three-dimensional object.)

    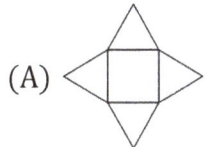 (A)   (B) 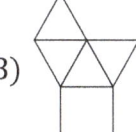 (C) 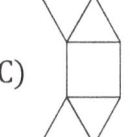 (D) 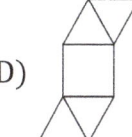 (E)

PROBLEMS 2015

14. Lucy and her mother were both born in February. On March 19, 2015, Lucy adds the year of her birth, the year of her mother's birth, her age, and her mother's age. What result does she get?

(A) 4028 (B) 4029 (C) 4030 (D) 4031 (E) 4032

15. A student wrote down a natural number. When she divided the number by 9, the remainder was 7. What is the remainder when twice that number is divided by 9?

(A) 1 (B) 2 (C) 5 (D) 6 (E) 7

16. The area of a rectangle is 12. The lengths of its sides are natural numbers. The perimeter of this rectangle could be:

(A) 20 (B) 26 (C) 32 (D) 28 (E) 24

17. Each of the segments in the figure needs to be colored one of three colors: red, green, or blue. Each of the 4 triangles needs to have each side of a different color. Three of the segments have already been colored, as indicated. What color can the segment marked with x be?

(A) only blue (B) only green (C) only red (D) any of the three colors
(E) Such a coloring is not possible.

18. In a bag there are 3 green apples, 5 yellow apples, 7 green pears, and 2 yellow pears. Simon randomly takes fruit out of the bag one by one. How many pieces of fruit must he take out in order to be sure that he has at least one apple and one pear of the same color?

(A) 9 (B) 10 (C) 11 (D) 12 (E) 13

19. A new chess piece called "kangaroo" has been introduced. In each move, it jumps either 3 squares vertically and 1 square horizontally, or 3 squares horizontally and 1 square vertically, as shown in the picture. What is the minimum number of moves the kangaroo needs to make in order to get from its current position to the square marked with A?

(A) 2 (B) 3 (C) 4 (D) 5 (E) 6

20. In this sum, the same letters represent the same digits and different letters represent different digits. Which digit does the letter X represent?

(A) 2 (B) 3 (C) 4 (D) 5 (E) 6

© Math Kangaroo in USA, NFP

PROBLEMS 2015

Problems 5 points each

21. The sum of four natural numbers is 39. The product of two of these numbers is equal to 80, and the product of the other two numbers is also equal to 80. What is the largest of these four numbers?

    (A) 8 (B) 10 (C) 16 (D) 20 (E) 25

22. Jane bought some books. For the first book, she paid half of her money and 1 dollar more. For the second book, she paid half of the remaining money and 2 dollars more. Finally, for the third book, she paid half of the remaining money and 3 dollars more, thus spending all of her money. How much did the three books cost altogether?

    (A) 36 dollars (B) 45 dollars (C) 34 dollars (D) 65 dollars (E) 100 dollars

23. On Jump Street there are 9 houses in a row. At least one person lives in each house. Any two neighboring houses together are inhabited by at most six people. What is the largest number of people that could be living on Jump Street?

    (A) 23 (B) 25 (C) 27 (D) 29 (E) 31

24. The number 100 is multiplied either by 2 or by 3, then the result is increased either by 1 or by 2, and then the new result is divided either by 3 or by 4. The final result is a natural number. What is this final result?

    (A) 50 (B) 51 (C) 67 (D) 68
    (E) There is more than one possible final result.

25. In a 4-digit number $\overline{abcd}$, the digits $a < b$, $b < c$, and $c < d$. What is the largest possible difference $\overline{bd} - \overline{ac}$ for 2-digit numbers $\overline{bd}$ and $\overline{ac}$?

    (A) 86 (B) 61 (C) 56 (D) 50 (E) 16

26. Mary wrote a number on each face of a cube. Then, for each vertex, she added the numbers on the three faces on which that vertex lies (for example, for vertex B she adds the numbers on faces BCDA, BAEF, and BFGC). She obtained 8 numbers. The numbers obtained by Mary for vertices C, D, and E are 14, 16, and 24, respectively. What number did she obtain for vertex F?

    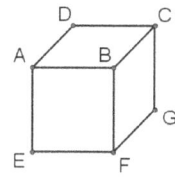

    (A) 15 (B) 19 (C) 22 (D) 24 (E) 26

27. Jane is traveling on a train in which each car has the same number of compartments. She is sitting in the 7th car, in the 50th compartment from the engine. How many compartments are there in each car?

    (A) 7 (B) 8 (C) 9 (D) 10 (E) 12

PROBLEMS 2015

28. Using only one color, in how many ways can you color 3 different cells in the strip shown below so that no 2 neighboring cells are colored?

   (A) 6   (B) 7   (C) 8   (D) 9   (E) 10

29. Four points lie on a line. The distances between them are, in increasing order: 2, 3, k, 11, 12, 14. What is the value of k?

   (A) 5   (B) 6   (C) 7   (D) 8   (E) 9

30. Basil used small cubes with a side length of 1 to construct a cube with a side length of 4. After that, he painted 3 faces of the big cube red and the other 3 faces blue. After he finished, there was no small cube with 3 red faces. How many small cubes have both red and blue faces?

   (A) 0   (B) 8   (C) 12   (D) 24   (E) 32

Problems from Year 2017

Problems 3 points each

1. Four cards lie in a row. | 2 | 0 | 1 | 7 |

   Which row of cards can you not obtain if you can only swap two cards?

   (A) 2 7 1 0   (B) 0 1 2 7   (C) 1 0 2 7
   (D) 0 2 1 7   (E) 2 0 7 1

2. A fly has 6 legs and a spider has 8 legs. Together, 3 flies and 2 spiders have as many legs as 9 chickens and

   (A) 2 cats.   (B) 3 cats.   (C) 4 cats.   (D) 5 cats.   (E) 6 cats.

3. Alice has 4 pieces of this shape: ▭. Which picture can she not make from these 4 pieces?

   (A)   (B)   (C)   (D)   (E)

4. Kalle knows that $1111 \times 1111 = 1234321$. How much is 1111×2222?

   (A) 3456543   (B) 2345432   (C) 2234322   (D) 2468642   (E) 4321234

PROBLEMS 2017

5. On a certain planet there are 10 islands and 12 bridges. All bridges are open for traffic right now. What is the smallest number of bridges that must be closed in order to stop all traffic between A and B?

   (A) 1   (B) 2   (C) 3   (D) 4   (E) 5

6. Jane, Kate, and Lynn go for a walk. Jane walks up front, Kate walks in the middle, and Lynn walks behind. Jane weighs 500 kg more than Kate. Kate weighs 1000 kg less than Lynn. Which of the following pictures shows Jane, Kate, and Lynn in the right order?

   (A)   (B)   (C)

   (D)   (E)

7. A special six-sided die has a number on each face. The sums of numbers on the opposite faces are all equal. Five of the numbers are 5, 6, 9, 11, and 14. What number is on the sixth face?

   (A) 4   (B) 7   (C) 8   (D) 13   (E) 15

8. Martin wants to color the squares of the rectangle so that ⅓ of all the squares are blue and half of all the squares are yellow. The rest of the squares are to be red. How many squares will he color red?

   (A) 1   (B) 2   (C) 3   (D) 4   (E) 5

9. In the time it takes Peter to solve 2 problems for the Math Kangaroo competition, Nick manages to solve three problems. In total, the boys solved 30 problems. How many more problems did Nick solve than Peter?

   (A) 5   (B) 6   (C) 7   (D) 8   (E) 9

10. Bob folded a piece of paper, then used a hole punch to punch exactly one hole in the paper.

    The unfolded paper can be seen in the picture: .

    Which of the following pictures shows the lines along which Bob folded this piece of paper?

    (A)   (B)   (C)   (D)   (E)

© Math Kangaroo in USA, NFP   www.mathkangaroo.org

Problems 4 points each

11. The Modern Furniture store is selling sofas, loveseats, and chairs made from identical modular pieces as shown in the picture. Including the armrests, the width of the sofa is 220 cm and the width of the loveseat is 160 cm. What is the width of the chair?

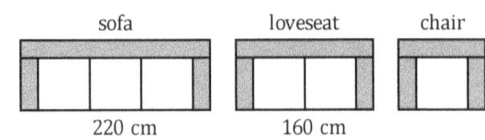

(A) 60 cm (B) 80 cm (C) 90 cm (D) 100 cm (E) 120 cm

12. The 5 keys fit the 5 padlocks. The numbers on the keys refer to the letters on the padlocks. What is written on the key with the question mark?

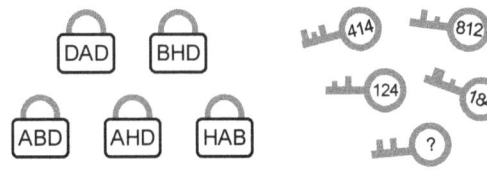

(A) 382 (B) 282 (C) 284 (D) 823 (E) 824

13. Tom writes all the numbers from 1 to 20 in a row and obtains the 31-digit number 1234567891011121314151617181920. Then he deletes 24 of the 31 digits in such a way that the remaining number is as large as possible. Which number does he get?

(A) 9671819 (B) 9567892 (C) 9781920 (D) 9912345 (E) 9818192

14. Morten wants to put the figure shown on the right into a regular box. Which of the following boxes is the smallest he can use?

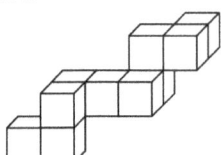

(A) $3 \times 3 \times 4$ (B) $3 \times 5 \times 5$ (C) $3 \times 4 \times 5$ (D) $4 \times 4 \times 4$ (E) $4 \times 4 \times 5$

15. When we add the numbers in each row and each column we get the results shown. Which statement is true?

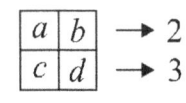

(A) a is equal to d. (B) b is equal to c. (C) a is greater than d.
(D) a is less than d. (E) c is greater than b.

16. Peter went hiking in the mountains for 5 days. He started on Monday and his last hike was on Friday. Each day he walked 2 km more than the day before. When the whole trip was over, his total distance was 70 km. What distance did Peter walk on Thursday?

(A) 12 km (B) 13 km (C) 14 km (D) 15 km (E) 16 km

PROBLEMS 2017

17. There is a picture of a kangaroo in the first triangle. A side shared by any two triangles acts as a mirror. The first 2 reflections are shown. What does the reflection look like in the shaded triangle?

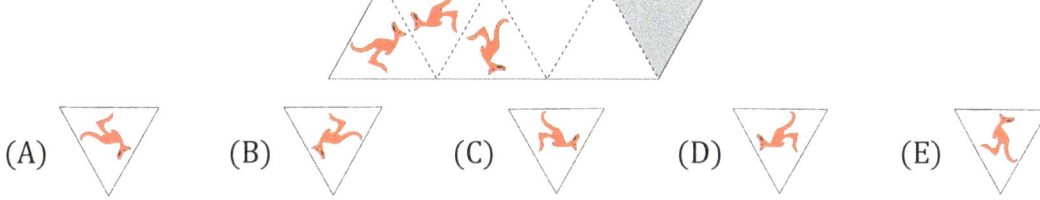

(A) 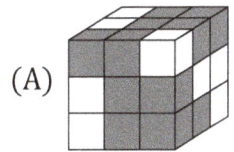 (B) 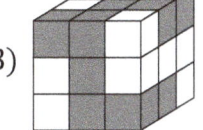 (C) 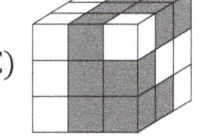 (D) 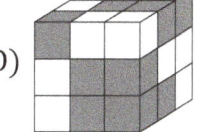 (E)

18. Boris has a certain amount of money and 3 magic wands that he can use only once.

This wand adds 1 dollar. This wand subtracts 1 dollar. This wand doubles the amount. In which order must he use these wands to obtain the largest amount of money?

(A) ×2 +1 −1   (B) +1 −1 ×2   (C) ×2 −1 +1   (D) +1 ×2 −1   (E) −1 +1 ×2

19. Rafael has three squares. The first one has a side length of 2 cm. The second one has a side length of 4 cm and a vertex placed in the center of the first square. The last one has a side length of 6 cm and a vertex placed in the center of the second square, as shown in the picture. What is the area of the entire shaded figure?

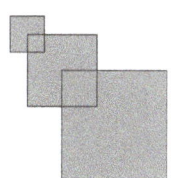

(A) 32 cm²   (B) 51 cm²   (C) 27 cm²   (D) 16 cm²   (E) 6 cm²

20. Four players scored goals in a handball match. All of them scored a different number of goals. Among the four, Mike was the one who scored the least number of goals. The other three scored 20 goals in total. What is the largest number of goals Mike could have scored?

(A) 2   (B) 3   (C) 4   (D) 5   (E) 6

Problems 5 points each

21. A bar consists of 2 gray cubes and 1 white cube glued together as shown here: 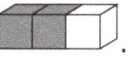. Which figure can be built from 9 such bars?

(A)   (B)   (C)   (D)   (E)

PROBLEMS 2017

22. The numbers 1, 2, 3, 4, and 5 have to be written in the five cells in the figure in the following way: If a number is just below another number, it has to be greater. If a number is just to the right of another number, it has to be greater. In how many ways can this be done?

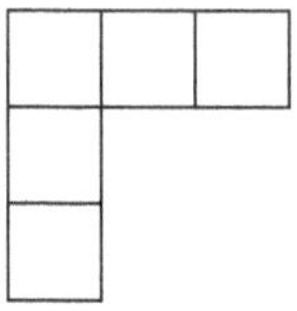

    (A) 3   (B) 4   (C) 5   (D) 6   (E) 8

23. 8 kangaroos stood in a line as shown in the diagram.

    At some point, two kangaroos standing side by side and facing each other exchanged places by jumping past each other. This was repeated until no further jumps were possible. How many exchanges were made?

    (A) 2   (B) 10   (C) 12   (D) 13   (E) 16

24. Monica has to choose 5 different numbers. She has to multiply some of the numbers by 2 and the remaining numbers by 3 in order to get the smallest number of different results. What is the least number of results she can obtain?

    (A) 1   (B) 2   (C) 3   (D) 4   (E) 5

25. The square floor in the picture is covered by triangular and square tiles in gray or white. What is the smallest number of gray tiles that need to be switched with white tiles so that the pattern looks the same from each of the four directions shown?

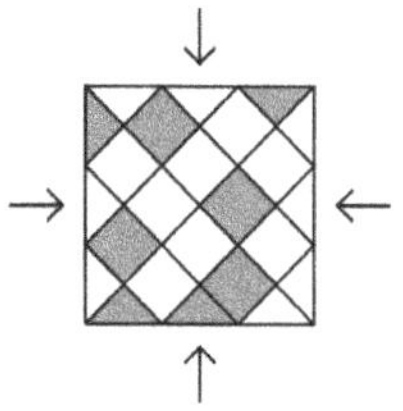

    (A) three triangles, one square   (B) one triangle, three squares
    (C) one triangle, one square   (D) three triangles, three squares
    (E) three triangles, two squares

26. A bag contains only red marbles and green marbles. For any 5 marbles we pick, at least one is red; for any 6 marbles we pick, at least one is green. What is the largest number of marbles that the bag can contain?

    (A) 11   (B) 10   (C) 9   (D) 8   (E) 7

© Math Kangaroo in USA, NFP   54   www.mathkangaroo.org

PROBLEMS 2017

27. Ala likes even numbers, Beata likes numbers divisible by 3, and Celina likes numbers divisible by 5. One by one, the girls walked up to a basket containing 8 balls with numbers written on them, and each girl took out all the balls with the numbers she likes. It turned out that Ala collected balls with numbers 32 and 52, Beata collected balls with numbers 24, 33, and 45, and Celina collected balls with numbers 20, 25, and 35. In what order did the girls approach the basket?

   (A) Ala, Celina, Beata   (B) Celina, Beata, Ala   (C) Beata, Ala, Celina
   (D) Beata, Celina, Ala   (E) Celina, Ala, Beata

28. John wants to write a natural number in each box in the diagram in such a way that each number above the bottom row is the sum of the two numbers in the boxes immediately underneath. What is the largest number of odd numbers that John can write?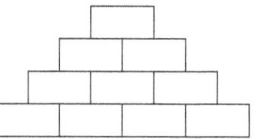

   (A) 4   (B) 5   (C) 6   (D) 7   (E) 8

29. Julia has four different colored pencils and wants to use some or all of them to color the map of an island divided into four countries, as shown in the picture. If two countries with a common border cannot be the same color, in how many ways can she color the map of the island?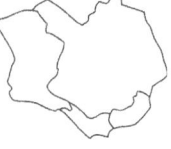

   (A) 12   (B) 18   (C) 24   (D) 36   (E) 48

30. In each cell of a 6 × 6 board there is a lamp. We say that two lamps on this board are neighbors if they lie in cells with a common side. At the beginning, some lamps are lit and each minute every lamp having at least two lit neighboring lamps is lit if the three cells have a common vertex. What is the minimum number of lamps that need to be lit at the beginning in order to be sure that at some time all lamps will be lit?

   (A) 4   (B) 5   (C) 6   (D) 7   (E) 8

Problems from Year 2019

Problems 3 points each

1. Carrie has started to draw a cat. She added more to her drawing using her black pen. Which of the figures below can be her finished drawing?

   (A)    (B)    (C)    (D)    (E)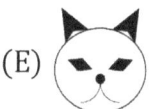

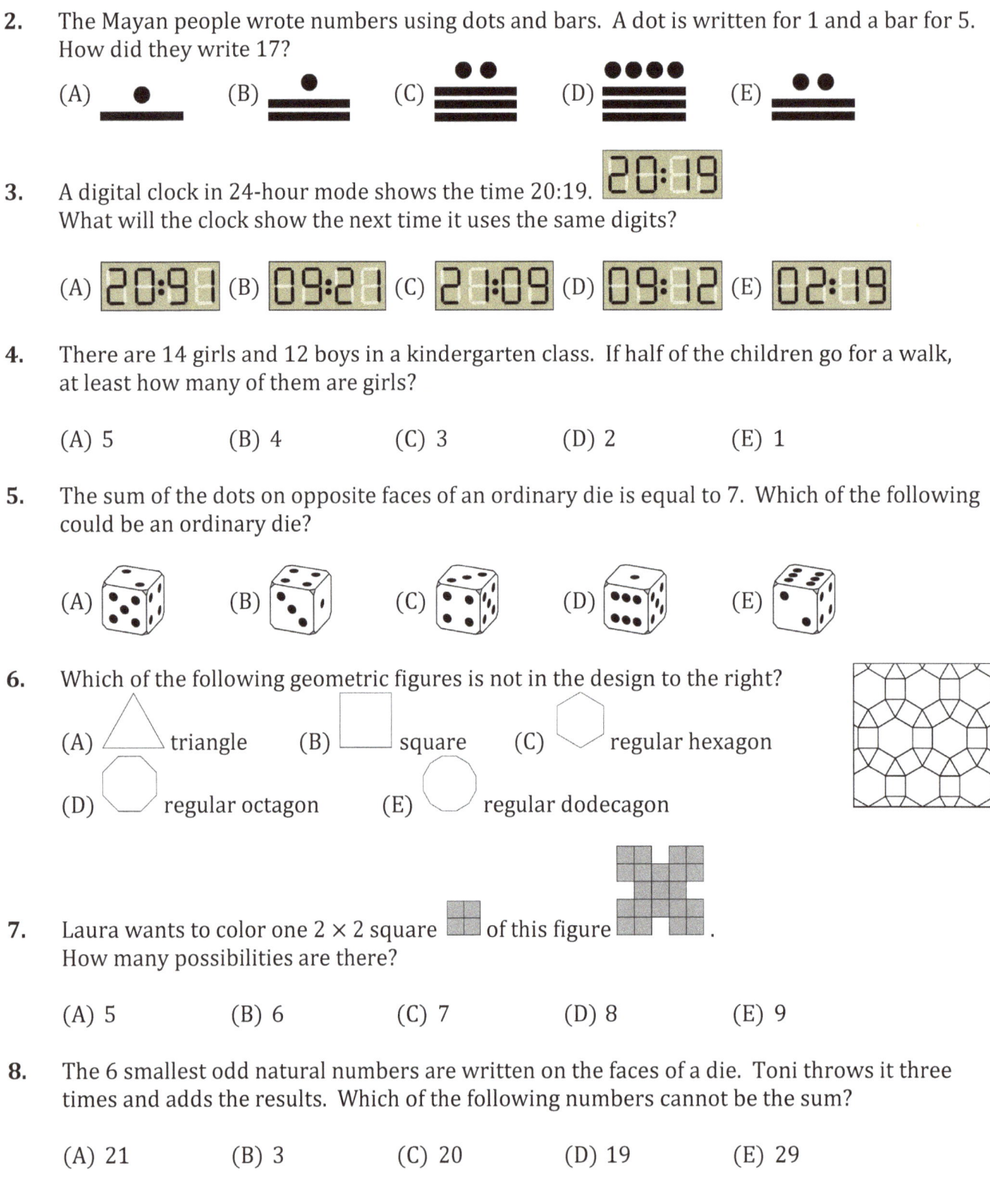

PROBLEMS 2019

9. The sum of the ages of a group of kangaroos is 36 years. In two years, the sum of their ages will be 60 years. How many kangaroos are in this group?

   (A) 10   (B) 12   (C) 15   (D) 20   (E) 24

10. Michael paints the following solids made out of identical cubes. Their bases are made of 8 cubes. Which solid needs the most paint?

    (A) 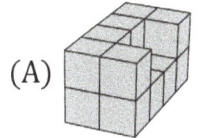   (B) 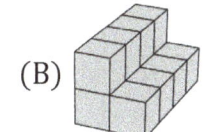   (C) 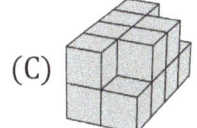   (D) 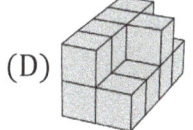   (E)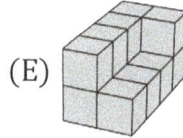

    Problems 4 points each

11. A three-digit number is written on each of three pieces of paper. Two of the digits are covered. The sum of the three numbers is 826. What is the sum of the two covered digits?

    (A) 7   (B) 8   (C) 9   (D) 10   (E) 11

12. Riri the frog usually eats 5 spiders a day. When Riri is very hungry, she eats 10 spiders a day. She ate 60 spiders in 9 days. On how many days was she very hungry?

    (A) 1   (B) 2   (C) 3   (D) 6   (E) 9

13. Pia is playing with a folding yardstick made of 10 parts (see picture). Which of the following figures cannot be formed with this folding yardstick?

    (A) 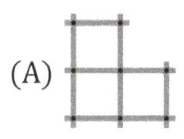   (B) 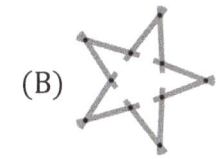   (C) 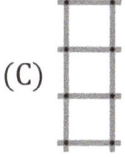   (D) 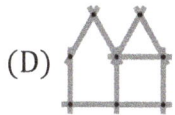   (E)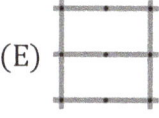

14. Five equal squares are divided into smaller squares. Which square has the largest black area?

    (A) 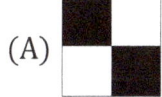   (B) 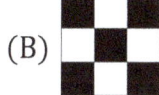   (C)    (D)    (E)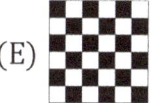

15. A big triangle is divided into equilateral triangles as shown in the figure. The side of the small gray triangle is 1 m. What is the perimeter of the big triangle?

    (A) 15 m   (B) 17 m   (C) 18 m   (D) 20 m   (E) 21 m

    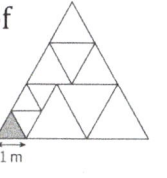

PROBLEMS 2019

16. In the garden of a witch there are 30 animals: dogs, cats, and mice. The witch turns 6 dogs into cats. Then she turns 5 cats into mice. Now the number of dogs, the number of cats, and the number of mice in her garden are all equal. How many cats were there at the beginning?

    (A) 4   (B) 5   (C) 9   (D) 10   (E) 11

17. With blocks of dimensions 1 cm × 1 cm × 2 cm, you can build towers as shown in the picture. How high is a tower that is built in the same way using 28 blocks?

    (A) 9 cm   (B) 11 cm   (C) 12 cm   (D) 14 cm   (E) 17 cm

18. Bridget folded a square sheet of paper twice, and then cut it twice as shown in the figure. How many pieces of paper did she get?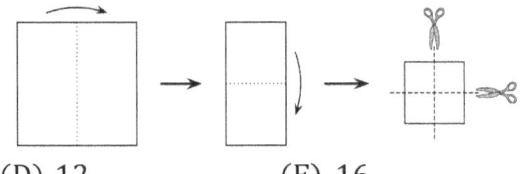

    (A) 6   (B) 8   (C) 9   (D) 12   (E) 16

19. Alex, Bob, and Carl go for a walk every day. If Alex doesn't wear a hat, then Bob wears a hat. If Bob doesn't wear a hat, then Carl wears a hat. Today Bob is not wearing a hat. Who is wearing a hat?

    (A) both Alex and Carl   (B) only Alex   (C) only Carl
    (D) neither Alex nor Carl   (E) It is not possible to determine.

20. Each of the following pictures shows the net of a cube. Only one of the resulting cubes has a line with connected endpoints drawn on it. Which one?

    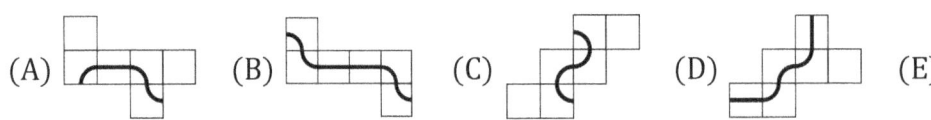

Problems 5 points each

21. The cube shown in the figure has a positive integer written on each face. The products of the two numbers on opposite faces are the same. What is the smallest possible sum of the six numbers on the cube?

    (A) 36   (B) 37   (C) 41   (D) 44   (E) 60

PROBLEMS 2019

22.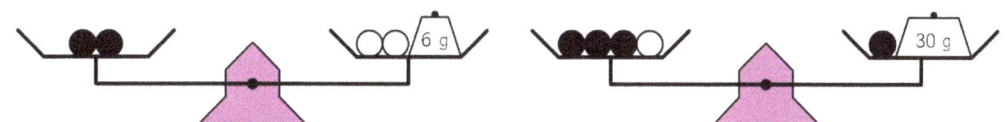

Six identical black beads and three identical white beads are arranged on scales as shown in the picture above. What is the total weight of these nine beads?

(A) 100 g (B) 99 g (C) 96 g (D) 94 g (E) 90 g

23. Robert made 5 statements (A) to (E). Exactly one of these statements is false. Which one?

(A) My son Basil has 3 sisters.   (B) My daughter Ann has 2 brothers.
(C) My daughter Ann has 2 sisters.   (D) My son Basil has 2 brothers.   (E) I have 5 children.

24. Benjamin writes an integer in the first circle and then fills the other five circles by following the instructions.

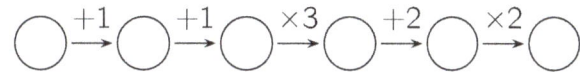

How many of the six numbers in the circles are divisible by 3?

(A) 1   (B) Both 1 and 2 are possible.   (C) 2
(D) Both 2 and 3 are possible.   (E) Both 3 and 4 are possible.

25. The cardboard shown on the right is folded into a 2 × 1 × 1 box. One of the pictures below does not show this box. Which one?

(A)    (B)    (C)    (D)    (E)

26. Emily took selfies with her 8 cousins. Each of the 8 cousins is in two or three pictures. In each picture there are exactly 5 of Emily's cousins. How many selfies did Emily take?

(A) 3   (B) 4   (C) 5   (D) 6   (E) 7

27. Jette and Willi are throwing balls at two identical pyramids of 15 cans. Jette knocks down 6 cans with a total of 25 points. Willi knocks down 4 cans. How many points does Willi score?

after Jette's throw   after Willi's throw

(A) 22   (B) 23   (C) 25   (D) 26   (E) 28

28. Every digit on my 24-hour digital clock is composed of at most 7 digital display parts, as follows:

But, unfortunately, for every digit display the same two digital parts are not working.

At this moment my clock shows: . What will it show after 3 hours and 45 minutes?

(A) 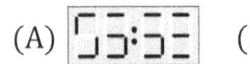 (B)  (C) 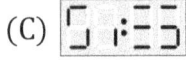 (D) 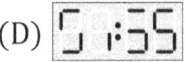 (E)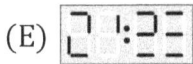

29. Linas builds a 4 × 4 × 4 cube using 32 white and 32 black 1 × 1 × 1 cubes. He arranges the cubes so that as much of the surface of his large cube as possible is white. What fraction of the surface of his cube is white?

(A) $\frac{1}{4}$ (B) $\frac{1}{2}$ (C) $\frac{2}{3}$ (D) $\frac{3}{4}$ (E) $\frac{3}{8}$

30. Zev has two machines: one makes 1 white token into 4 red tokens, and the other makes 1 red token into 3 white tokens. Zev starts with 4 white tokens. After exactly 11 exchanges, he has 31 tokens. How many of those are red?

(A) 21   (B) 17   (C) 14   (D) 27   (E) 11

Problems from Year 2021

Problems 3 points each

1. Which of the following solid shapes can be made with these 6 bricks?

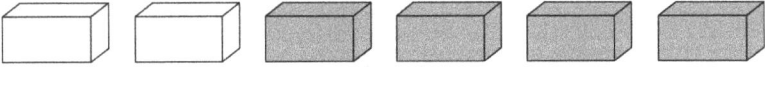

(A) 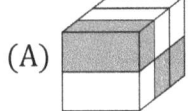 (B) 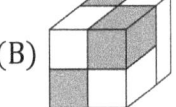 (C) 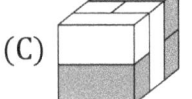 (D) 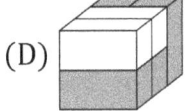 (E)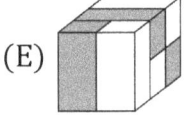

PROBLEMS 2021

2. In how many places in the picture are two children holding each other with their left hands?

(A) 1 (B) 2 (C) 3 (D) 4 (E) 5

3. In the square you can see the digits from 1 to 9. A number is created by starting at the star, following the line and writing down the digits along the line while passing. For example the line shown to the right represents the number 42685. Which of the following lines represents the largest number?

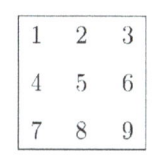 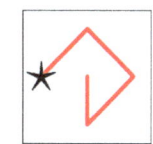

(A) 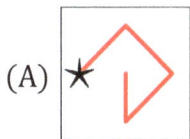 (B)  (C)  (D) 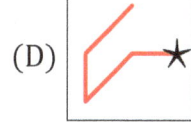 (E)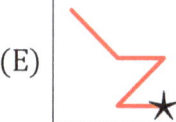

4. Sofie wants to write the word KENGU by using letters from the boxes. She can only take one letter from each box. What letter must Sofie take from box 4?

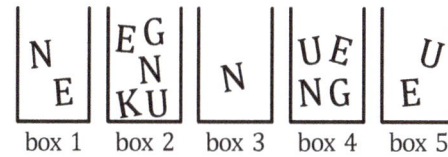

(A) K (B) E (C) N (D) G (E) U

5. When the 5 pieces shown fit together correctly, the result is a rectangle with a calculation written on it. What is the answer to this calculation?

(A) 22 (B) 32 (C) 41 (D) 122 (E) 203

6. A measuring tape is wrapped around a cylinder. Which number should be at the place shown by the question mark?

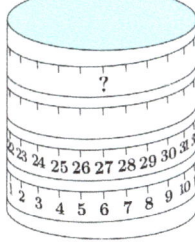

(A) 53 (B) 60 (C) 69 (D) 77 (E) 81

© Math Kangaroo in USA, NFP www.mathkangaroo.org

PROBLEMS 2021

7. The 5 figures on the grid can move only in the directions indicated by the black arrows. Which figure can leave through gate G?

   (A) A   (B) B   (C) C   (D) D   (E) E

   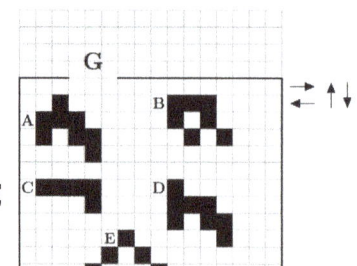

8. Carin is going to paint the walls of her room green. The green paint is too dark so she mixes it with white paint. She tries different mixtures. Which of the following mixtures will give the darkest green color?

   (A) 1 part green + 3 parts white   (B) 2 parts green + 6 parts white
   (C) 3 parts green + 9 parts white   (D) 4 part green + 12 parts white
   (E) They will all be equally dark.

9. Mary had a piece of paper. She folded it exactly in half. Then she folded it exactly in half again.
   She got this shape 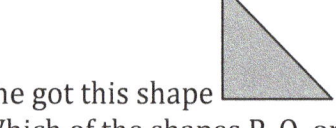 .
   Which of the shapes P, Q, or R could have been the shape of her original piece of paper?

   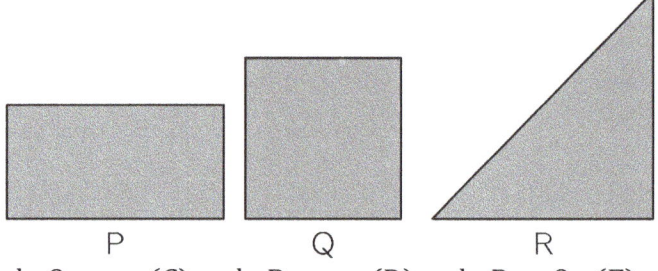

   (A) only P   (B) only Q   (C) only R   (D) only P or Q   (E) any of P, Q, or R

10. There is a square with line segments drawn inside it. The line segments are drawn either from the vertices or the midpoints of other line segments. We colored $\frac{1}{8}$ of the large square. Which one is our coloring?

    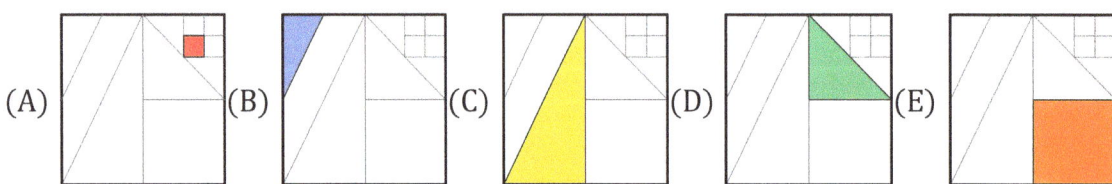

PROBLEMS 2021

Problems 4 points each

11. The number 5021972970 is written on a sheet of paper. Julian cuts the sheet twice so he gets three numbers. What is the smallest sum he can get by adding these three numbers?

   (A) 3244   (B) 3444   (C) 5172   (D) 5217   (E) 5444

12. The map shows three bus stations at points A, B and C. A tour from station A to the Zoo and the Port and back to A is 10 km long. A tour from station B to the Park and the Zoo and back to B is 12 km long. A tour from station C to the Port and the Park and back to C is 13 km long. Also, a tour from the Zoo to the Park and the Port and back to the Zoo is 15 km long. How long is the shortest tour from A to B to C and back to A?

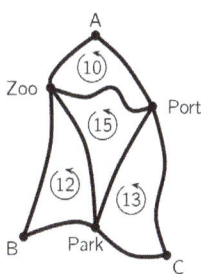

   (A) 18 km   (B) 20 km   (C) 25 km   (D) 35 km   (E) 50 km

13. Rosa wants to start at the arrow, follow the line, and get out at the other arrow. Which piece, if placed in the middle, cannot produce this? Note: The pieces can be rotated.

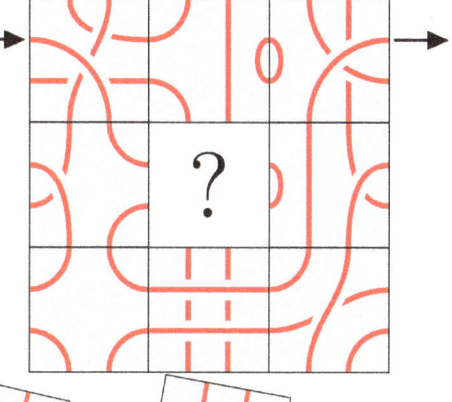

   (A) 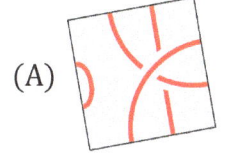   (B) 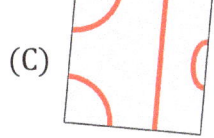   (C)   (D)   (E)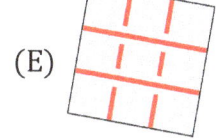

14. The diagram shows three hexagons with numbers at their vertices, but some numbers are invisible. The sum of the six numbers around each hexagon is 30. What is the number on the vertex marked with a question mark?

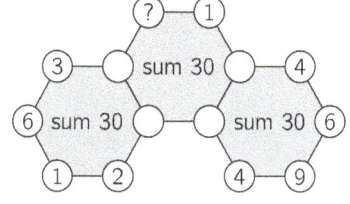

   (A) 3   (B) 4   (C) 5   (D) 6   (E) 7

15. Three rectangles of the same height are positioned as shown. The numbers within the rectangles indicate their areas in cm². If AB = 6 cm, how long is CD?

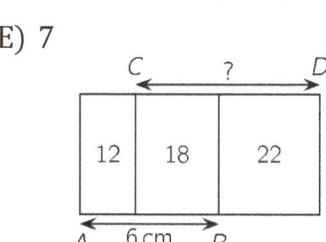

   (A) 7 cm   (B) 7.5 cm   (C) 8 cm   (D) 8.2 cm   (E) 8.5 cm

PROBLEMS 2021

16. A triangular pyramid is built with 10 identical balls, as shown. Each ball has one of the letters A, B, C, D, and E on it. There are 2 balls marked with each letter. The picture shows three side views of the pyramid. What is the letter on the ball with the question mark?

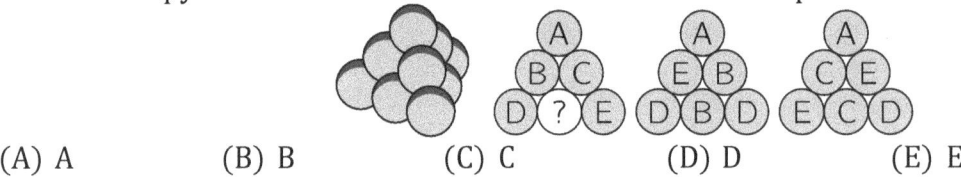

(A) A   (B) B   (C) C   (D) D   (E) E

17. Ronja had four white tokens and Wanja had four gray tokens. They played a game in which they took turns to place one of their tokens to create two piles. Ronja placed her first token first. Which pair of piles could they not have created?

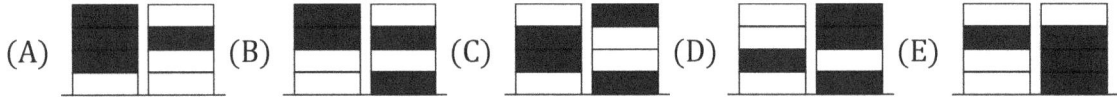

18. My little brother has a 4-digit bike lock with the digits 0 to 9 on each part of the lock as shown. He started on the correct combination and turned each part the same amount in the same direction and now the lock shows the combination 6348. Which of the following CANNOT be the correct combination of my brother's lock?

19. There were 20 apples and 20 pears in a box. Carl randomly took 20 pieces of fruit from the box and Luca took the rest. Which of the following statements is always true?

(A) Carl got at least one pear.   (B) Carl got as many apples as pears.
(C) Carl got as many apples as Luca.   (D) Carl got as many pears as Luca got apples.
(E) Carl got as many pears as Luca.

20. There is a single train track between points X and Y. A train company wants one train to leave from X and one train to leave from Y at the same time daily. Moving with constant speed it takes 180 minutes for a train to make a trip from X to Y and 60 minutes from Y to X. They want to build a double track to avoid a crash. Where should the double track be?

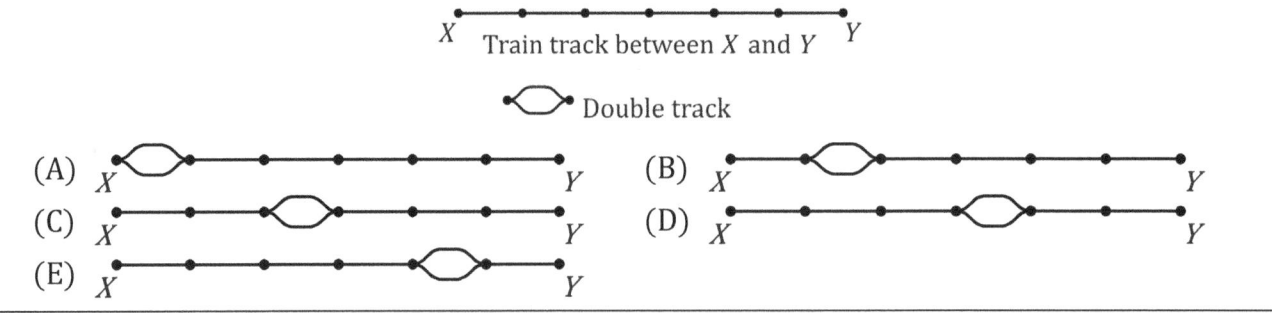

PROBLEMS 2021

Problems 5 points each

21. Ann, Bob, Carina, Dan, and Ed are sitting at a round table. Ann is not next to Bob, Dan is next to Ed, and Bob is not next to Dan. Which two people are sitting next to Carina?

(A) Ann and Bob   (B) Bob and Dan   (C) Dan and Ed   (D) Ed and Ann
(E) It is not possible to be certain.

22. Maurice asked the canteen chef for the recipe for his pancakes. Maurice has 6 eggs, 400 g of flour, 0.5 liters of milk, and 200 g of butter. What is the largest number of pancakes he can make using this recipe (see the picture) and only the ingredients he has?

Ingredients for 100 pancakes
25 eggs   4 l milk
5 kg flour   1 kg butter

(A) 6   (B) 8   (C) 10   (D) 12   (E) 15

23. The picture shows three gears with a black gear tooth on each. Which picture shows the correct position of the black teeth after the small gear has turned a full turn clockwise?

 (A)    (B)    (C)    (D)    (E)

24. An apple and an orange weigh as much as a pear and a peach weigh together. An apple and a pear weigh less than an orange and a peach together, and a pear and an orange weigh less than an apple and a peach together. Which of the pieces of fruit is the heaviest?

(A) apple   (B) orange   (C) peach   (D) pear
(E) It is impossible to determine.

25. What is the smallest number of colored squares that can be added so that this 6×6 square has four axes of symmetry?

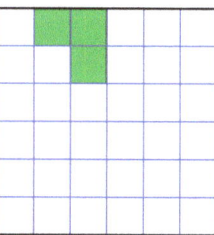

(A) 1   (B) 9   (C) 12   (D) 13   (E) 21

26. Three pirates were asked how many coins and how many diamonds their friend Graybeard had. Each of the three answered truthfully to one question but lied answering the other. Their answers are written on the piece of paper pictured. What is the total number of coins and diamonds that Graybeard has?

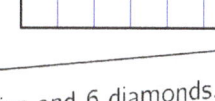

(1) He has 8 coins and 6 diamonds.
(2) He has 7 coins and 4 diamonds.
(3) He has 7 coins and 7 diamonds.

(A) 11   (B) 12   (C) 13   (D) 14   (E) 15

PROBLEMS 2021

27. Each shelf holds a total of 64 deciliters of apple juice. The bottles have three different sizes: large, medium, and small. How many deciliters of apple juice does a medium bottle contain?

(A) 3   (B) 6   (C) 8   (D) 10   (E) 14

28. A large cube has an edge with a length of 7 cm. On each of its 6 faces, the two diagonals are drawn in red. The large cube is then cut into small cubes with edges 1 cm long. How many small cubes will have at least one red line drawn on it?

(A) 54   (B) 62   (C) 70   (D) 78   (E) 86

29. 10 elves and trolls each were given a token with a different number from 1 to 10 written on it. They were each asked what number was on their token and all answered with a number from 1 to 10. The sum of the answers was 36. Each troll told a lie and each elf told the truth. What is the smallest number of trolls there could be in the group?

(A) 1   (B) 3   (C) 4   (D) 5   (E) 7

30. There are rectangular cards divided into four equal cells with different shapes □, ✯, ●, ▲ drawn in each cell. Cards can be placed side by side only if the same shapes appear in adjacent cells on their common side. Nine cards are used to form a rectangle as shown in the figure. Which of the following cards was definitely NOT used to form this rectangle?

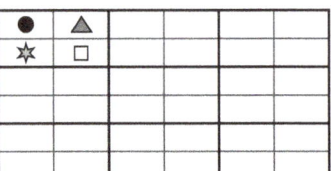

(A)    (B)    (C)    (D)    (E)

Part II: Solutions

Solutions for Year 1999

1. (B) 1090
    Do the operations left to right.
    1999 − 999 + 99 − 9 =
    = 1000 + 99 − 9 =
    = 1099 − 9 =
    = 1090

2. (D) Both ways have the same length.

    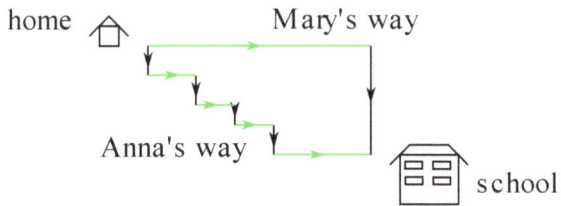
    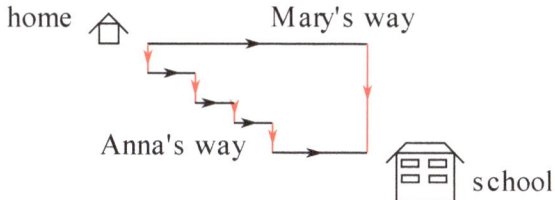

    Ann's horizontal segments (red) match Ann's vertical segments (red) match
    Mary's horizontal segment (green).   Mary's vertical segment (green).
    Therefore, both ways have the same length.

3. (B) 8
    $\frac{1}{4} \times \left[\frac{1}{2} \times (2 \times 32)\right] = \frac{1}{4} \times 32 = 8$

4. (D) 6
    There are 3 ways from **A** to **B** (the left upper arc, the left straight segment, and the left lower arc) and 2 ways from **B** to **C** (the right upper arc and the right lower arc), so $3 \times 2 = 6$ different ways that connect city **A** to city **C** (any left part can be connected with any right part).

5. (B) 12
    The minute hand makes one full revolution during 1 hour and the hour hand makes one full revolution during 12 hours, so the minute hand moves 12 times faster than the hour hand.

6. (D) 9
    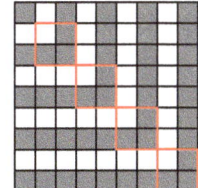
    To the **left** of each red frame the shaded unit squares match the white unit squares. The same is true for the unit squares **above** each red frame, so without the upper left unit square (shaded) and without unit squares inside red frames the shaded unit squares match the white unit squares. In each red frame there are 2 more shaded squares than white squares, so in the 4 red frames there are $4 \times 2 = 8$ more shaded squares than white squares. Add the upper left shaded unit square to see that $4 \times 2 + 1 = 8 + 1 = 9$ is the difference between the number of shaded squares and white squares.

SOLUTIONS 1999

7. (D) 36 cm²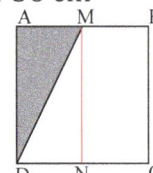

MN is perpendicular to AB and DC and M is the midpoint of AB, so the shaded area is $\frac{1}{4}$ of the square ABCD. The shaded part has an area equal to 9 cm², so the area of the square ABCD is 4×9 cm² = 36 cm².

8. (B) 151 min

The time from 1:47 p.m. to 4:18 p.m. is 13 min + 2 hours + 18 min = 2 h 31 min, which is 2×60 min + 31 min or 151 min, so the movie was 151 min long.

9. (B) 12

There are 12 different months, so the greatest possible number of people at the party is 12, just one birthday each month.

10. (A) 26

The empty spaces are filled with same size pink bricks. Count them row by row to see that there are $1 + 3 + 5 + 10 + 3 + 4 = 26$ pink bricks, so 26 bricks were taken from the wall.

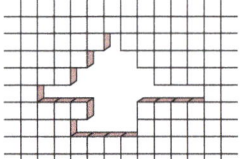 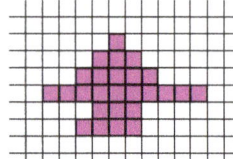

11. (D) 156

The numbers on the two pages of the book are consecutive natural numbers. Since their sum is 25, the numbers are 12 and 13. Their product is $12 \times 13 = 156$.

12. (D) 1111

1 m = 10 dm = 100 cm = 1000 mm
5000 m + 5000 dm + 5000 cm + 5000 mm = 5000 m + 500 m + 50 m + 5 m = 5555 m. The kangaroo's jump is 5 m long and $5555 \div 5 = 1111$, so the kangaroo needs 1111 jumps to cover the distance of 5000 m + 5000 dm + 5000 cm + 5000 mm.

13. (B) 225

1 m in reality is represented by 2 cm in the picture.
4.5 cm = 2.25×2 cm, so the actual height of the fence is 2.25×1 m = 2.25 m = 225 cm.

14. (E) 14

    6A3
  ×   5
  346B

The given product ends with 5 since $3 \times 5 = 15$ ends with 5, so the product is 3465.
The number 5 is one of the factors, so the other factor is $3465 \div 5 = 693$.
The missing digits are 9 and 5 and their sum is $9 + 5 = 14$.

SOLUTIONS 1999

15. (C)

As you can see in the picture below, only figure C will form a rectangle when combined with the figure on the right.

(A) (B) (C) (D) (E)

16. (A) 30
The dog's weight is 9 times the cat's weight, which is 20 times the mouse's weight, so the dog's weight is 180 times the mouse's weight since 9 × 20 = 180. 180 = 30 × 6 and the turnip's weight is 6 times the mouse's weight, so the dog is 30 times heavier than the turnip.

17. (C) 126
1.5 m = 150 cm = 15 × 10 cm and 1 m = 100 cm = 10 × 10 cm, so the quilt will have 15 columns and 10 rows of square scraps. The numbers of rows and columns where the squares meet (the seams) will be one less than the numbers of columns and rows of squares. Thus, there will be 14 columns of buttons with 9 buttons in each column, so the kangaroo will need 9 × 14 = 126 buttons.

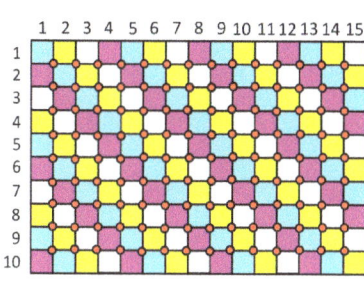

18. (A) 2.5
Each blue strip represents 7 liters of water, and each yellow strip represents the amount of
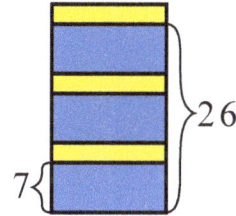
water added to the bowl with 7 liters of water. It is the same as the amount of water added to the bowl with 26 liters of water. The bowl with 7 liters of water and the additional amount of water is shown 3 times and it is the same as 26 liters of water with the additional amount of water. Thus, 3 blue strips and 2 yellow strips represent 26 liters of water. Hence, 1 yellow strip represents (26 − 3 × 7) ÷ 2 = 5 ÷ 2 = 2.5 liters of water. Therefore, 2.5 liters of water was added to each bowl.

19. (D) 25

16	3	A
C	10	
B		4

The magic square shown to the left has only one solution. The sum of numbers along one diagonal is 16 + 10 + 4 = 30. This is the same sum as in each of the rows and each of the columns.
The number covered with A is 30 − (16 + 3) = 30 − 19 = 11. In the first column, the sum is 16 + B + C = 30, so B + C = 30 − 16 = 14. We don't need to solve further to find that A + B + C = 11 + 14 = 25.

If we do want to solve the whole square, the last entry in the 2nd column is 30 − (3 + 10) = 17. After that the middle entry in the 3rd column is computed as 30 − (11 + 4) = 15. Then, the first entries of the second and third row are computed as 30 − (10 + 15) = 5 and 30 − (17 + 4) = 9.

16	3	11
5	10	15
9	17	4

SOLUTIONS 1999

20. (A) 32

The number of squares is: 8 green, 16 yellow, 24 blue, and 32 orange. Thus, Adam needs to add 32 squares (the orange ones) to the fourth square to make the fifth one. Each time, the number of squares of the new color increases by 8 compared to the previous color. To see it, look at the four 2 × 2 corner squares with black perimeters. Without the unit squares inside them, the orange squares match the blue squares. In each of the four 2 × 2 squares there are 3 orange and 1 blue square, so there are 4 × (3 − 1) = 8 more orange squares than the blue ones.

Another solution: The fifth square is a 9 × 9 square and the fourth square is a 7 × 7 square, so the number of new unit squares (in orange) is $9^2 − 7^2 = 32$.

21. (D) 8

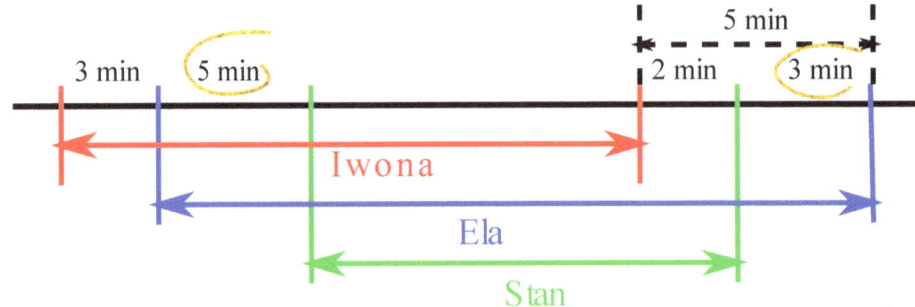

Ela arrived at the birthday party 5 minutes before Stan. Iwona left the party 2 minutes before Stan and 5 minutes before Ela, so Ela left the party 3 minutes after Stan. Hence, Ela was at the party 5 + 3 = 8 minutes longer than Stan.

22. (C) 6006

For each group of 606 people we need 6 hotdogs without mustard.
606,606 = 606,000 + 606 = 1000 × 606 + 606 = 1001 × 606, so for 606,606 people we will need 1001 × 6 = 6006 hotdogs without mustard.

23. (D) 634

Either the second or the third squirrel ate **at least** 633 nuts. Otherwise, each of them would eat 632 nuts or less, so together they would eat not more than 632 + 632 = 1264 nuts, which is less than the 1265 nuts actually eaten by the two squirrels. The first squirrel ate more nuts than any other squirrel, so the first squirrel ate **at least** 634 nuts. Together the first and the fourth squirrel ate 1999 − 1265 = 734 nuts. The fourth squirrel alone ate 100 or more nuts, so the first squirrel ate **not more** than 734 − 100 = 634 nuts but we already know that the first squirrel ate **at least** 634 nuts. Therefore, the first squirrel ate exactly 634 nuts. Consequently, the fourth squirrel ate exactly 734 − 634 = 100 nuts. Of the other two squirrels, one ate exactly 633 nuts and another of them ate exactly 632 nuts.

24. (D) The cat is in the basement and the mouse is in the room.
Right now, the cheese is on the table, so **the cat is not in the room** because the cat in the room means that the cheese is in the refrigerator.
Thus, the cat is in the basement and the cheese is on the table. It means that the mouse is in the room. Therefore, the cat is in the basement and the mouse is in the room. This is the only option that is true. Each of the other options is false.

25. (E) 15

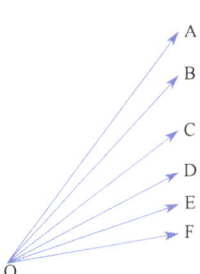

$\angle AOF$ and all other angles are acute angles (less than 90°), so we can pick any two rays to have an acute angle. Here is the list of all te angles.
$\angle AOB$, $\angle AOC$, $\angle AOD$, $\angle AOE$, $\angle AOF$, $\angle BOC$, $\angle BOD$, $\angle BOE$, $\angle BOF$, $\angle COD$, $\angle COE$, $\angle COF$, $\angle DOE$, $\angle DOF$, and $\angle EOF$. There are 15 such angles.
For any collection of 6 items, 15 is the number of all pairs that can be formed by using only these 6 items. If we call the item A, B, C, D, E, and F, then the possible pairs are: AB, AC, AD, AE, AF, BC, BD, BE, BF, CD, CE, CF, DE, DF, and EF.

26. (B) 15
It seems that the "yield" for any natural number written in the decimal system is the product of its digits, so yield (30) = 3 × 0 = 0, yield (36) = 3 × 6 = 18, yield (45) = 4 × 5 = 20, and yield (325) = 3 × 2 × 5 = 30. If this is the case, then yield (531) = 5 × 3 × 1 = 15.

27. (E)

When folding the net into a cube, keep the face with N as the front face, the face with M as the left face, the face with R as the top face, the face with S as the bottom face, the face with O as the right face, and the face with P as the back face. The faces with R and S, the faces with N and P, and the faces with M and O are pairs of opposite faces which eliminates

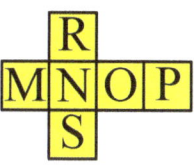

cubes (A), (B), and (D)  as cubes of the net.

In (C) the positions of the letters M and N with respect to the common edge are incorrect. Two parts (legs) of M and two parts (legs) of N are perpendicular to the common edge of the two faces but in the net these parts are parallel to the common edge. This eliminates cube (C) as a cube of the net. (E) is the only option left.

SOLUTIONS 1999

28. (B)

Each cup contains only one beverage, so there is exactly one beverage that is only in 1 cup and each of the other two beverages is in 2 cups (1 + 2 + 2 = 5). The problem asks which cup the cocoa is in, implying that the cocoa is the beverage that is in one cup. If this is the case, the tea and coffee are each in two cups.

The one cup of cocoa must match half of the coffee in two other cups. From 5 given cups we can select two cups in 10 ways. Add the numbers of the two selected cups and divide the sum by 2. If the result matches a number of another cup, then the cup matched contains cocoa.
(950 + 750) ÷ 2 = 850; (950 + 550) ÷ 2 = 750; (950 + 475) ÷ 2 = 712.5;
(950 + 325) ÷ 2 = 637.5; (750 + 550) ÷ 2 = 650; (750 + 475) ÷ 2 = 612.5;
(750 + 325) ÷ 2 = 537.5; (550 + 475) ÷ 2 = 512.5; (550 + 325) ÷ 2 = 437.5;
and (475 + 325) ÷ 2 = 400.

Only selecting the 950 g and 550 g cups for coffee produces a match. The cocoa is in the 750 g cup, the coffee is in the 950 g and 550 g cups, and tea is in the 475 g and 325 g cups.

The amount of coffee is 1500 g, which is twice as much as the amount of cocoa.

29. (B) 47

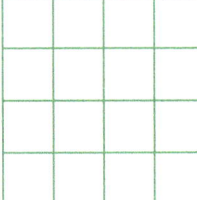

In the figure to the left there are squares of sizes 1 × 1, 2 × 2, 3 × 3, and 4 × 4. To get a square of the size 1 × 1 you need a pair of vertical lines 1 unit apart (there are 4 such pairs) and a pair of horizontal lines 1 unit apart (there are 4 such pairs). Thus, there are $4^2 = 16$ squares of the size 1 × 1. To get a square of the size 2 × 2 you need a pair of vertical lines 2 units apart (there are 3 such pairs) and a pair of horizontal lines 2 units apart (there are 3 such pairs). Thus, there are $3^2 = 9$ squares of the size 2 × 2.
To get a square of the size 3 × 3 you need a pair of vertical lines 3 units apart (there are 2 such pairs) and a pair of horizontal lines 3 units apart (there are 2 such pairs). Thus, there are $2^2 = 4$ squares of the size 3 × 3.
To get a square of the size 4 × 4 you need a pair of vertical lines 4 units apart (there is 1 such pair) and a pair of horizontal lines 4 units apart (there is 1 such pair). Thus, there is $1^2 = 1$ square of the size 4 × 4.
Therefore, there are $1^2 + 2^2 + 3^2 + 4^2 = 30$ squares in the 4 × 4 square above.

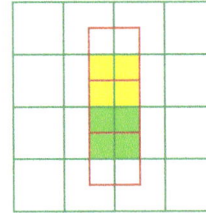

3 squares (with red perimeters) are added to the above figure. In addition to these 3 squares, there are 2 more 1 × 1 new squares (shown in colors). There are also 2 × 6 = 12 very small new squares. Hence, 3 + 2 + 12 = 17 new squares have been added to the 30 squares. Altogether, there are 30 + 17 = 47 squares.

SOLUTIONS 1999

30. **(E) 898**
    Right now the electronic clock shows 19:58:47. In one second it will show 19:58:48, which contains two 8s, and one second later it will show 19:58:49, which contains two 9s. Afterwards the digit 5 will be repeated multiple times and then 9 will be repeated for one minute. Once we reach 8 PM, 20:00:00, we will see 20:13:45 as the first electronic display with all different digits. The time from 19:58:47 to 20:13:45 is 13 sec + 1 min + 13 min + 45 sec or 14 min + 58 sec = 14 × 60 sec + 58 sec = 898 seconds.

Solutions for Year 2001

1. **(B) 0**
   Do multiplication before addition, so $2 \times 0 + 0 \times 1 = 0 + 0 = 0$.

2. **(D) $\frac{1}{12}$**
   One box represents $\frac{1}{6}$ of the rectangle, so the shaded area represents $\frac{1}{2} \times \frac{1}{6} = \frac{1}{12}$ of the rectangle.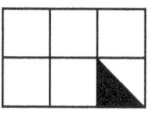

3. **(C)**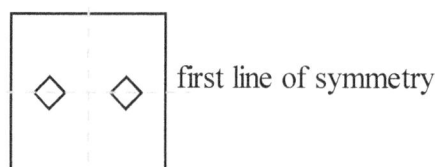
   The top edge of the folded napkin is one line of symmetry. After the first unfolding, the edges on the left become one edge which is also a line of symmetry. In terms of the first line of symmetry, after the first unfolding the paper cut-out looks like a diamond. Then, after unfolding along the second line of symmetry, we see two diamonds. After rotating the unfolded napkin by 90° we get the napkin shown in (C).

4. **(B) 8 minutes**
   24 × 20 seconds = 480 seconds = 8 × 60 seconds = 8 minutes, so after 24 hours the clock will be late 8 minutes.

5. **(E) 72**
   The proportion of occupied seats to the empty seats is 2 : 1, so the proportion of occupied seats to all seats is 2 : (2 + 1). The number of passengers on the plane was
   $\frac{2}{2+1} \times 108 = \frac{2}{3} \times 108 = 2 \times 36 = 72$

6. **(C) 12**
   Ella has (3 − 1) sisters (excluding herself) and (5 + 1) brothers (including Johnny). The product is 2 × 6 = 12.

7. **(D) 84**
Simon doubled his original number 4 times, so the result of this operation is
2 × 2 × 2 × 2 × the original number. The result is a multiple of 16 since 2 × 2 × 2 × 2 = 16.
On the list only 84 is not a multiple of 16, so it cannot be the result of Simon's operation.
80 = 5 × 16, 1200 = 75 × 16, 48 = 3 × 16, and 880 = 55 × 16.

8. (E)

Notice that two shaded triangles cover exactly one shaded square, and a shaded circular quadrant (a quarter of a circle) covers more than one shaded triangle.
Let s be the area of one small square. Let c be the area of one circular quadrant.
Then $\frac{1}{2}s < c < s$ (or $s < 2c < 2s$). For each of the regions below we will compute the area in terms of s and c.

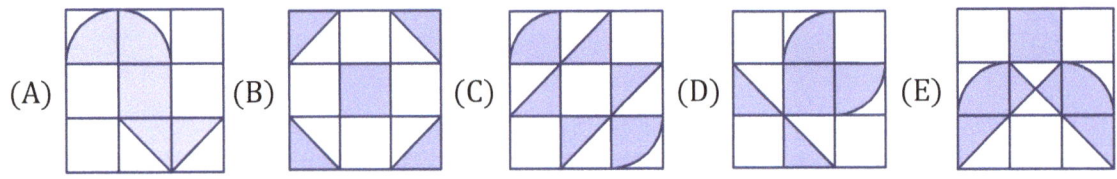

For (A) the area is $\left(1 + \frac{1}{2} + \frac{1}{2}\right)s + (1+1)c = 2s + 2c$.
For (B) the area is $\left(\frac{1}{2} + \frac{1}{2} + 1 + \frac{1}{2} + \frac{1}{2}\right)s = 3s$.
For (C) the area is $\left(\frac{1}{2} + \frac{1}{2} + \frac{1}{2} + \frac{1}{2}\right)s + (1+1)c = 2s + 2c$.
For (D) the area is $\left(\frac{1}{2} + 1 + \frac{1}{2}\right)s + (1+1)c = 2s + 2c$.
For (E) the area is $\left(1 + \frac{1}{4} + \frac{1}{4} + \frac{1}{2} + \frac{1}{2}\right)s + (1+1)c = 2\frac{1}{2}s + 2c$.

Notice that $3s = 2s + s < 2s + 2c$ and $2s + 2c < 2\frac{1}{2}s + 2c$, so (B) has the smallest shaded area and (E) has the largest shaded area.

9. **(C) 2641**
According to Picture 1 and Picture 2, there is no circle around ones digit, there is one circle around tens digit, and two circles around hundreds digit. Following the pattern, we will see 3 circles around thousands digit.

Thus, 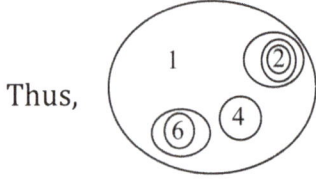 represents the number 2641.

SOLUTIONS 2001

10. (A) 2
There are 8 squares visible in the original picture, 3 are large (3 × 3), 2 are medium (2 × 2), and 3 are small (1 × 1). By adding the two matches shown here in pink, **three** more small squares will be added for a total of 11 squares altogether.

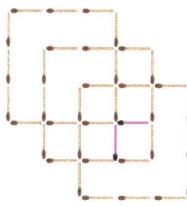

11. (D) 12
The least common multiple of 3 and 4 is 12, so the boys will meet again after 12 minutes.

12. (A) 536 dollars
$201 \div 3 = 67$, so Anya has sixty seven 1 dollar coins, sixty seven 2 dollar coins, and sixty seven 5 dollar coins. Altogether, she has $67 \times (1 + 2 + 5)$ dollars, which is 536 dollars.

13. (C) 900
8,631 meters + 3,456 decimeters + 12,340 centimeters is
8,631 meters + 345.6 meters + 123.40 meters = 9,100 meters.
10 kilometers – 9,100 meters is 10,000 meters – 9,100 meters = 900 meters, so Adam had 900 meters left to finish the race.

14. (C) 3 cm
The seven sticks, each of the length 14 centimeters, cover the length of 80 centimeters. There are 6 overlaps (indicated by "?") of equal length, so the length of one overlap is $(7 \times 14 \text{ cm} - 80 \text{ cm}) \div 6 = 18 \text{ cm} \div 6 = 3 \text{ cm}$.

15. (B) 8
The green dragon and any other dragon with 6 more heads together have 34 heads, so 2 identical green dragons would have $34 - 6 = 28$ heads. Hence, the green dragon has 14 heads. The red dragon has $(14 - 6)$ heads, which is 8 heads.

16. (D) 80 cm
A given rectangle has the length ℓ, the width w, and the area A. Another rectangle has the length ℓ', the width $\frac{w}{2}$, and the area $\frac{A}{2}$, so $\ell \times w = A$ and $\ell' \times \frac{w}{2} = \frac{A}{2}$.
The last equation is equivalent to $\ell' \times w = A$. $\ell \times w = A$, so $\ell' = \ell$. The original rectangle has a length of 80 cm, so the new rectangle also has a length of 80 cm.
Or: The area of a rectangle is found by multiplying the length by the width. If the width is reduced by one-half and the length stays the same, then the area is reduced by one-half. So, the second rectangle has the same length as the first rectangle, that is, 80 cm.

17. (C) 24
1 hour is 60 minutes, so Zosia spends 20 minutes doing math. She spends two fifths of the remaining 40 minutes working on geography, so she works $\frac{2}{5} \times 40$ minutes = 16 minutes on geography. 20 minutes + 16 minutes = 36 minutes and $60 - 36 = 24$, so she spends 24 minutes doing the rest of her homework.

SOLUTIONS 2001

18. **(D) 12**

If you add 3 years + 3 years + 3 years + 3 years to 24 years, then you see that today the four siblings are altogether 36 years old. To compensate for the age difference between Mary and the triplets add 4 years + 4 years + 4 years to 36 years to see Mary's age being quadrupled. Hence, 4 × Mary's age = 48 years. 48 ÷ 4 = 12, so Mary is 12 years old today.

19. **(E) 900**

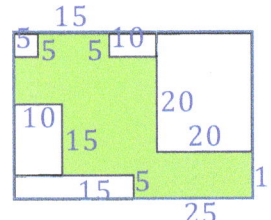

Extend the garden to a full rectangle by adding rectangular parcels as shown in the picture. The top edge of the extended rectangle has length 5 + 15 + 10 + 20 = 50. The right edge of the extended rectangle has length 5 + 20 + 10 = 35, so the area of the extended rectangle is 35 × 50 = 1,750.
The areas of the added rectangular parcels (counted along the top edge and then along the left edge) are: 5 × 5, 5 × 10, (5 + 20) × 20, 15 × 10, and 5 × (10 + 15). Their sum is 25 + 50 + 500 + 150 + 125 = 850.
1,750 − 850 = 900, so the area of Jonah's garden is 900 square meters.

20. **(B) 40**

Adam's money is 4 times Charlie's money and Bart's money is 2 times Charlie's money, so the three of them together earned 4 + 2 + 1 times Charlie's money. Therefore, Charlie earned 280 ÷ 7 = 40 dollars.

21. **(A) 1**

7 × 7 × 7 ends with the digit 3, so 7 × 7 × 7 × 7 = (7 × 7 × 7) × 7 ends with the last digit of 3 × 7, which is 1. $\underbrace{7 \times 7 \times 7 \times 7 \times 7 \times 7 \times 7 \times 7 \times ... \times 7 \times 7 \times 7 \times 7}_{100\ times}$ is $\underbrace{(7 \times 7 \times 7 \times 7) \times (7 \times 7 \times 7 \times 7) \times ... \times (7 \times 7 \times 7 \times 7)}_{25\ times}$, so its last digit is the last digit of $\underbrace{1 \times 1 \times ... \times 1}_{25\ times}$, which is 1.

22. **(B) 34**

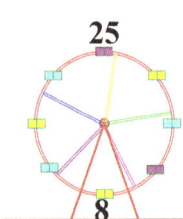

Moving clockwise from the cabin with number 8 to the cabin with number 25, we encounter cabins with the following numbers: 9, 10, 11, 12, 13, 14, 15, 16, 17, 18, 19, 20, 21, 22, 23, and 24. There are 16 of them. By symmetry, there are also 16 cabins when we move counterclockwise from 8 to 25. Together, not counting the cabins with numbers 8 and 25, there are 32 cabins. Including those two cabins, the Big Wheel has 34 cabins.

SOLUTIONS 2001

23. (A)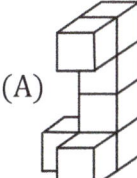
Two cubes are adjacent if they share a face. For a given cube, the number of its faces to be painted depends on the number of cubes adjacent to the given cube. So, for a given cube the number of faces to be painted is 6 minus the number of adjacent cubes. For each solid we count (from top to bottom, left to right, front to back) the number of faces to be painted.
(A) 5 + 4 + 4 + 4 + 5 + 3 + 5 = 30,   (B) 5 + 5 + 3 + 3 + 4 + 4 + 4 = 28,
(C) 4 + 4 + 5 + 2 + 3 + 4 + 4 = 26,   (D) 5 + 4 + 3 + 4 + 3 + 4 + 5 = 28, and
(E) 5 + 3 + 4 + 3 + 4 + (invisible) 3 + 4 = 26.
Thus, the solid (A) requires the largest amount of paint.

24. (C) 10 cm²
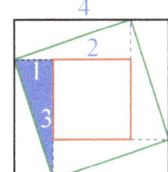
Each side of the biggest square is 4 cm, since 4 cm × 4 cm = 16 cm². Each side of the smallest square is 2 cm, since 2 cm × 2 cm = 4 cm². So, the legs of the shaded right triangle are 1 cm and 3 cm. The medium square consists of four such triangles and the smallest square, so the area of the medium square is 4 × (½ × 3 × 1) cm² + 4 cm², which is 10 cm².

25. (E) 96
Each die, without counting the top and bottom faces of the solid, contributes two pairs of opposite faces to the surface of the solid. The numbers of dots on each pair of opposite faces add up to 7; so, each die (without counting the top and bottom faces of the solid) contributes 14 dots to the surface of the solid. Without counting the top and bottom faces of the solid, the six dice together contribute 6 × 14 = 84 dots to the surface of the solid.
To have the largest number of dots on the surface we need 6 dots on the top face and 6 dots on the bottom face, so the largest number of dots on the whole surface is 84 + 6 + 6, which is 96.

26. (D) more than 21
The number 3___ is between 3,000 and 3,999. $3{,}000 \div 45 = 200 \div 3 = 66 + \frac{2}{3}$ and $3{,}999 \div 45 = 1{,}333 \div 15 = 88 + \frac{13}{15}$, so $67 \leq \_3 \leq 88$ since _3 is a natural number and $45 \times \_3 = 3\_\_\_$. Natural numbers between 67 and 88 with the last digit 3 are 73 and 83. In the first case, 45 × 73 = **3285**. The sum of the digits in the blanks is **7 + 2 + 8 + 5**, which is 22. In the second case, 45 × 83 = **3735**. The sum of the digits in the blanks is **8 + 7 + 3 + 5**, which is 23. In either case, the sum is not 20, it is not 21, it is not 17, and it is not less than 17, but it is more than 21.

© Math Kangaroo in USA, NFP   www.mathkangaroo.org

SOLUTIONS 2001

27. (A) 88

The small cubes along the edges of the big cube are not affected by tunnels. Among these cubes there are 8 corner cubes since any cube has 8 corners. The other small cubes along the edges do not overlap and along each edge there are (**5** – 2) = 3 such cubes. A cube always has 3 × 4 = 12 edges (three directions and 4 edges in each direction), so the total number of small cubes along the edges is 8 + 12 × (**5** – 2) = 44.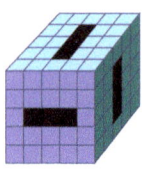
In the top layer of the big cube there are still 2 × 3 = 6 small cubes (along the tunnel opening) not affected by the tunnels and not counted yet. The same applies to any of the 6 outer layers of the big cube and these small cubes do not overlap. Therefore, the number of all small cubes of the six outer layers which are not affected by the tunnels is 44 + 6 × (2 × 3) = 44 + 36 = = 80. Take away these 80 cubes from the big cube. You are left with a 3 × 3 × 3 cube and three mutually perpendicular tunnels through the center of the 3 × 3 × 3 cube. Only the 8 corners of the 3 × 3 × 3 cube are not affected by the tunnels, so the total number of small cubes left in the solid is 80 + 8 = 88.

28. (C) 14

In one hour the students consume (34 × 0.7) kg of oxygen, so the number of oaks we need is (34 × 0.7) kg ÷ 1.7 kg. $\frac{34 \times 0.7}{1.7} = \frac{34 \times 7}{17} = 14$, so 14 such oak trees are needed.

29. (C) 12

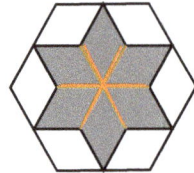 Draw 3 line segments through the center of the hexagon, parallel to the sides of the hexagon, as shown in the picture. The whole hexagon is divided into 12 congruent rhombuses, 6 of them shaded and 6 of them white. Hence the shaded region of the hexagon (the star) is half the area of the hexagon. Therefore, the area of the hexagon is 2 × 6 = 12.

30. (D) 47

Let *abc* and *def* be 3-digit numbers such that *a, b, c, d, e, f* (in some order) are the digits 1, 2, 3, 4, 5, and 6, and *a* < *d*. The difference is 100 × (*d* – *a*) + 10 × (*e* – *b*) + (*f* – *c*).
Intuitively, keep *a* and *d* as close as possible with *a* < *d*, keep *b* and *e* as far apart as possible with *e* < *b*, and keep *f* and *c* also distant with *f* < *c* but respect the other requirements. Thus, *a* and *d* should be selected form the middle of the list 1, 2, 3, 4, 5, 6 and *e* and *b* from the start and the end of the list.
365 and 412 are such numbers and the difference is 412 – 365 = 47.
47 is the actual answer.
Indeed, if (*d* – *a*) ≥ 2, then the difference is greater than 100.
If (*d* – *a*) = 1 but (*e* – *b*) is not –5, then the difference is greater than 50.
Thus, we need (*d* – *a*) = 1 and (*e* – *b*) = –5 , so *e* = 1 and *b* = 6 are necessary for the smallest difference. After that *f* = 2 and *c* = 5 is the optimal option left, so *a* must be 3 and *d* must be 4. Clearly, 412 – 365 = 47 is the smallest possible difference.

Solutions for Year 2003

1. (C) $(2 + 0) \times (0 + 3)$
    One factor equal to 0 makes the whole product equal to 0 and adding 0 to a number doesn't change the number, so $2 \times 0 \times 0 \times 3$, $20 \times 0 \times 3$, and $(2 \times 0) + (0 \times 3)$ are all equal to 0. $2 + 0 + 0 + 3 = 5$ and $(2 + 0) \times (0 + 3) = 2 \times 3 = 6$, so $(2 + 0) \times (0 + 3)$ is the greatest number among the five given numbers.

2. (A) blue
    Zosia is using 4 different colors. $29 = 7 \times 4 + 1$, so the twenty ninth flower drawn by Zosia has the same color as the first one, which is blue.

3. (A) 13
    Integers between 2.09 and 15.3 are 3, 4, 5, ..., 14, 15. There are $15 - 2 = 13$ of them.

4. (C) 12
    The least positive integer divisible by 2 and 3 is 6. 6 is not divisible by 4. The next multiple of 6 is 12, and 12 is the least positive integer divisible by 2, 3, and 4.

5. (B) 10
    $B + 9 + 9 + A + 8 + 11 = 55$ and $B + 11 + 14 + 2 + 13 + 7 = 55$, so
    $B + A + 37 = B + 47$. Hence, $A = 10$. Notice that for computing A we only
    have to know that $B + 9 + 9 + A + 8 + 11 = B + 11 + 14 + 2 + 13 + 7$.
    If both sums are equal to 55, then $B = 8$.

    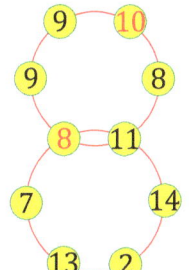

6. (A) $1,000
    $9 \times \$100 + 9 \times \$10 + 10 \times \$1 = \$900 + \$90 + \$10 = \$1000$, so Tom has $1,000.

7. (E) 11 cm

    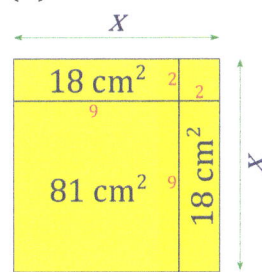

    Each side of the large square is 9 cm since $(9 \text{ cm})^2 = 81 \text{ cm}^2$.
    The rectangle above the large square has one side equal to 9 cm and
    an area of 18 cm², so the vertical side of the rectangle is equal to 2 cm
    since $2 \text{ cm} \times 9 \text{ cm} = 18 \text{ cm}^2$. Therefore, the left side of the big square
    is 9 cm + 2 cm, which is 11 cm. Hence, $x = 11$ cm.
    Notice that the area of the small square at the top right corner is 4 cm²
    and $81 + 2 \times 18 + 4 = 9^2 + 2 \times 9 \times 2 + 2^2 = (9 + 2)^2 = 11^2$.

8. (D) $\frac{5}{2}$

    $$\frac{2003 + 2003 + 2003 + 2003 + 2003}{2003 + 2003} = \frac{5 \times 2003}{2 \times 2003} = \frac{5}{2}$$

SOLUTIONS 2003

9. (A) 24

The first minute after midnight is reported as 00:01. Then 00:02, ...,00:59, 01:00, 01:01, ... 01:59, 02:00, 02:01,..., 02:59, 3:00, ..., 23:00, 23:01,..., 23:59.
After that, the time is either 24:00, 00:01, 00:02,... or 00:00, 00:01, 00:02,.... There is no 24:01 or 24:59. The sum of digits for 24:00 is 6 and the sum of digits for 23:59 is 19 but 19 is not the greatest sum. It is true that $5 + 9 = 14$ is the greatest minute sum but the greatest hour sum is $1 + 9 = 10$, so the greatest sum of all four digits is $10 + 14 = 24$. It is the sum for 19:59.

10. (C) 22 dm

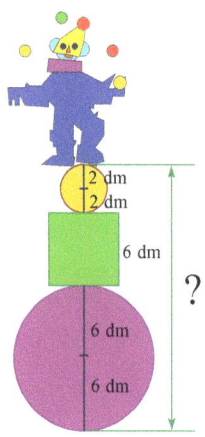

A radius is a segment from the center of a circle or sphere to its circumference or surface. The diameter is the maximum distance between points on the circumference or surface.
The radius of the lower ball is 6 dm and the radius of the upper ball is 2 dm since $6 \div 3 = 2$. The edge of the cube is 2 dm + 4 dm = 6 dm. Ian is dancing at a height which is the sum of the diameters of both balls and the edge of the cube. $2 \times 6 + 6 + 2 \times 2 = 22$, so Ian is dancing at the height of 22 dm.

11. (C) 3 m

AD is 22 meters long and BD is 15 meters long, so AB is $22 - 15 = 7$ meters long. AC is 10 meters long and AB is 7 meters long, so BC is $10 - 7 = 3$ meters long. Hence, the length of segment BC is equal to 3 m.

12. (B) 6

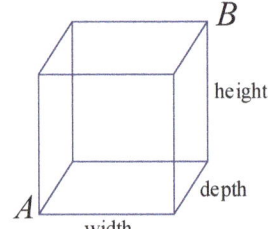

From each vertex we can move in 3 directions: width, depth, height. The shortest path consists of three consecutive edges in 3 different directions selected in any order. There are 6 such options: w-d-h, w-h-d, d-w-h, d-h-w, h-w-d, and h-d-w.

13. (D) 1 and 3

The shaded areas is made up of 17 small squares.
Pieces 1 and 2 together cover 10 + 9 = 19 small squares, so they cannot fit in the shaded area.
Pieces 1 and 4 together cover 10 + 8 = 18 small squares, so they cannot fit in the shaded area. Pieces 2 and 3 together cover 9 + 7 = 16 small squares, so they cannot fill the shaded area. Also, pieces 3 and 4 together cover 7 + 8 = 15 small squares, so they cannot fill the shaded area. The only pairs of pieces that cover an area are pieces 1 and 3 (10 + 7 = 17) and pieces 2 and 4 (9 + 8 = 17).

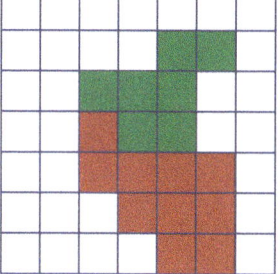

After rotating 90° clockwise (as seen to the left), piece 3 fits in the green area and piece 1 fits directly in the red area.

The only two ways that piece 4 can fit into the shaded region is shown in orange below. There is no possibility to fit any other pieces to cover the remaining shaded area.

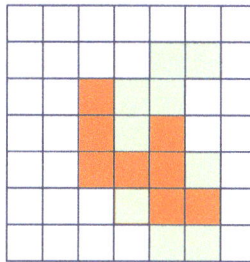 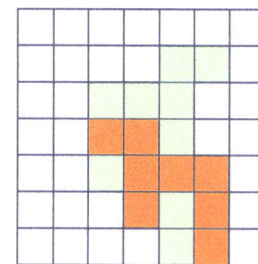

Therefore, only the combination of Pieces 1 and 3 is the right fit for the shaded region.

14. (C) 7

The smallest sum is 1 + 2 = 3 and the largest sum is 4 + 5 = 9. There are 7 integers from 3 to 9, and each of them is one of the sums. Indeed, 4 = 1 + 3, 5 = 1 + 4, 6 = 1 + 5, 7 = 2 + 5, and 8 = 3 + 5. Hence, we can get 7 different sums.

15. (B) 25

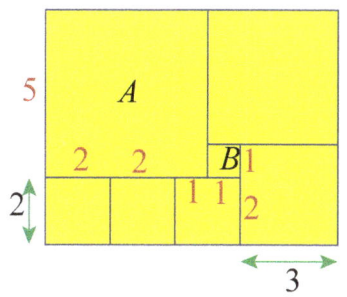

The difference between the left edge of 3 × 3 square (the bottom right corner of the big rectangle) and the right edge of 2 × 2 square adjacent to it is 1, so B is a 1 × 1 square. Hence the bottom edge of the square A is 2 + 2 + 2 − 1 = 5. A is a 5 × 5 square and B is a 1 × 1 square, so we need 25 B-squares to fill in square A completely.

SOLUTIONS 2003

16. (D) 4
There are 16 white bars since the 1st bar is black and the 17th bar is black. There are 13 wide black bars since 16 – 3 = 13. Also, 17 – 13 = 4, so there are 4 narrow black bars in this bar code.

17. (B) 4
Eva has 3 green marbles since 20 – 17 = 3. She has 8 yellow marbles since 20 – 12 = 8. 5 marbles are black, so Eva has 20 – (3 + 8 + 5) = 4 blue marbles.

18. (B) 5
Enumerate consecutive trees from Tom's house to school as
1, 2, 3, 4, 5, 6, 7, 8, 9, 10, 11, 12, 13, 14, 15, 16, 17. When he walks to school, he marks trees 1, 3, 5, 7, 9, 11, 13, 15, 17. When he walks back, he marks trees 17, 14, 11, 8, 5, and 2.
The trees 4, 6, 10, 12, and 16 are not marked, and there are 5 of them. Notice that 4, 6, 10, 12, 16 are even numbers between 1 and 17 such that (17 – even number) is not a multiple of 3.

19. (C) 3.23.2003
2003 minutes = 33 × 60 minutes + 23 minutes = 33 hours + 23 minutes =
= 24 hours + 9 hours + 23 minutes = 1 day + 9 hours + 23 minutes. Today the time is 20:03, so one day later will be tomorrow (March 22nd) at 20:03; 9 hours later will be the day after tomorrow (March 23rd) at 5:03 a.m., and 23 minutes later will be the day after tomorrow at 5:26. Thus, the date after 2003 minutes will be 3.23.2003 (March 23rd, 2003, at 5:26 a.m.).

20. (A) 7
The last digit of 2003^4 is the last digit of $3 \times 3 \times 3 \times 3 = 9 \times 9 = 81$, so the last digit of 2003^4 is 1. $2003^{2003} = [(2003)^4]^{500} \times 2003^3$, so the last digit of 2003^{2003} is the last digit of 2003^3. The last digit of 2003^3 is the last digit of $3 \times 3 \times 3 = 27$, which is 7. Hence, the last digit of 2003^{2003} is 7.

21. (C) 12
5, 5 × 2, 5 × 3, 5 × 4, 5 × 5, 5 × 6, 5 × 7, 5 × 8, 5 × 9, and 5 × 5 × 2 are multiples of 5 from 1 to 50. The factor 5 is repeated 12 times in the prime factorization of the product of consecutive numbers from 1 to 50. The factor 2 is repeated more than 12 times (actually 47 times) but we only need to pair up 2s with the twelve 5s to get twelve factors of 10. Therefore, the product of consecutive numbers from 1 to 50 ends with 12 zeros.

22. (D) 400 cm²
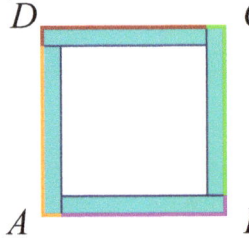
Half of the perimeter of each shaded rectangle is 20 cm and each shaded rectangle shares half of its perimeter with the perimeter of the square *ABCD* as shown in four different colors for the four rectangles. These four halves cover exactly the perimeter of the square *ABCD*, so the perimeter of the square is 4 × 20 cm. Hence, each side of the square has the length of (4 × 20 cm) ÷ 4 = 20 cm and the area of the square *ABCD* is (20 cm)², which is 400 cm².

SOLUTIONS 2003

23. (D) 6

In any triangle the sum of the lengths of any two sides is strictly greater than the length of the third side. It is enough when the sum of two shorter lengths is greater than the longest one. This immediately excludes 1 as the shortest side.
2 can be the shortest side in **two** triangles since $2 + 2001 > 2002$ and $2 + 2002 > 2003$.
3 can be the shortest side in **three** triangles since $3 + 2001 > 2002$, $3 + 2001 > 2003$, and $3 + 2002 > 2003$. Also, $2001 + 2002 > 2003$. Thus, we can build 6 triangles using the given segments.

24. (B)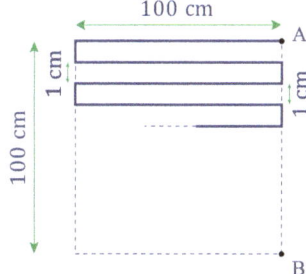

In each column the last digits are two alternating digits. The two alternating digits are: 0 and 1 in the first column, 2 and 3 in the second column, 4 and 5 in the third column, 6 and 7 in the fourth column, and 8 and 9 in the fifth column. Thus, there can be no cell to the right of the column with 8 or 9. Hence, the piece (B) cannot be a part of the table. To the right, the table is extended to confirm that the other pieces shown below are parts of the table.

25. (D) 10,100 cm

Enumerate vertical segments (dashed, solid, dashed, solid, and so on) from A to B along the right edge. The odd numbers indicate the dashed segments and the even numbers indicate the solid segments. The last segment (ending at B) is the 100th segment (100 is an even number), so it is solid. This vertical 1 cm long segment is the last segment of the path from A to B, so the last horizontal segment of the path is just 1 cm above the bottom edge. Therefore, the path contains 100 horizontal segments each 100 cm long. They add up to 100×100 cm $= 10{,}000$ cm. The vertical dashed segments have their counterparts along the left edge as segments of the path, so all the vertical pieces of the path add up to 100 cm. $10{,}000$ cm $+ 100$ cm $= 10{,}100$ cm, so the whole path is 10,100 cm long.

26. (E) 35°

The minute hand makes a full rotation (360°) in 60 minutes, so in 10 minutes it moves 60° to 2 on the face of the clock. The angle between two consecutive numbers marked on the clock is $360° \div 12 = 30°$. Thus, the hour hand moves 30° in 1 hour. 1 hour = 60 minutes, so during 10 minutes the hour hand moves $30° \div 6 = 5°$.

The angle from "2" to "3" is 30°, and the hour hand is 5° beyond "3," so at 3:10 the angle between the hands is $30° + 5° = 35°$.

SOLUTIONS 2003

27. (A) 6

If you insert 5 (or less) in the squares, then you are adding 3 numbers less than 600 each. The sum cannot be 2003 since it is less than 1800. If you insert 7 (or more) in the squares, then you are adding 3 numbers greater than (or equal to) 700 each. The sum cannot be 2003 since it is at least 2100. Hence, the square represents 6. 2003 is the sum of the three numbers, so 2003 = 666 + 660 + the circle digit + 600 + the two-digit number with identical digits. The circle digit + the two-digit number with identical digits = 2003 − (666 + 660 + 600) = = 2003 − 1926 = 77. Any two-digit number with identical digits is a multiple of 11, so the circle digit = (77 − the two-digit number with identical digits) is a multiple of 11. The only digit that is a multiple of 11 is 0, so the circle digit is 0 and the number represented by two triangle digits is 77. 666 + 660 + 677 = 2003 and it is the only solution for this puzzle, so the sum of the numbers represented by the square and by the circle is 6 + 0 = 6.

28. (D) 45 cm²

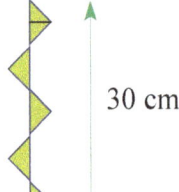

The hypotenuse of each triangle is equal to 30 cm ÷ 5 = 6 cm. The height of the triangle (shown) divides it into two isosceles right triangles, so the height is equal to half of the original hypotenuse. Each shaded triangle has the base equal to 6 cm and the height equal to 3 cm. The area of one shaded triangle is ½ × 6 cm × 3 cm = 9 cm², so the area of the shaded figure is 5 × 9 cm² = 45 cm².

29. (C) 8

Notation: r is the number of red dragons and g is the number of green dragons.
The red dragons have 6r heads, 8r legs, and 2r tails.
The green dragons have 8g heads, 6g legs, and 4g tails.
2r + 4g = 44 and 6g = 6r − 6. From the second equation g = r − 1. Substitute it into the first equation to have 2r + 4(r − 1) = 44. Simplify it to 6r = 44 + 4. 6r = 48, so r = 8.
Thus, 8 red dragons (and 7 green dragons) lived in the cave.

30. (B) 3

There are at most 3 crayon colors. Otherwise, there is a group of 4 crayons in 4 different colors which contradicts, "At least two of every 4 crayons are of the same color." There are at most 3 crayons in each color. Otherwise, there is a group of 4 crayons in one color. Add any crayon to this group to contradict, "At most three out of every 5 crayons are of the same color." This means that there are at most 3 blue crayons.
If there are only two (or fewer) blue crayons, then the total number of all crayons is at most 2 (blue) + 3 + 3 (at most two other colors; not more than 3 crayons in each color). 2 + 3 + 3 = 8 but the total number is 9. Therefore, there are exactly 3 blue crayons in the box. In fact, there are exactly 3 crayon colors and exactly 3 crayons in each color.

Solutions for Year 2005

1. (E) 1290
   $2005 + 205 = 2210$
   $3500 - 2210 = 1290$
   So, the butterfly covered the number 1290.

2. (C) 6
   Olla and Anna will have equal numbers of candy if Anna gets two more pieces. Together they will have 12 pieces of candy, half of them Olla's, so Olla has 6 pieces of candy.

3. (B) 1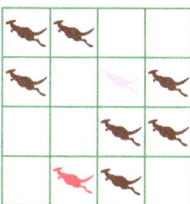
   The 2nd row and the 3rd column have 3 kangaroos each but the 4th row and the 2nd column have 1 kangaroo each. Move the kangaroo from the 2nd row of the 3rd column to the 4th row of the 2nd column to have two kangaroos in each row and each column.

4. (D) 24
   Eva, her brother, and their parents together have 8 legs. A dog and two cats together have 12 legs, two parrots have 4 legs, and four goldfish have no legs. Altogether they have
   $8 + 12 + 14 = 24$ legs.

5. (D) 202505
   $2005 \times 100 + 2005 = 200500 + 2005 = 202505$

6. (D) 60 cm
   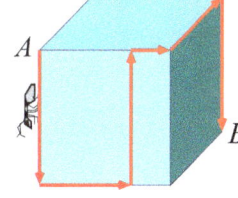
   There are 3 vertical arrows of the length of one edge each. Two shorter arrows, along the edges of the front face, add up to one edge, and there is one full arrow along the edge of the top face. Thus, the ant's path is equivalent to the length of 5 edges, which is 5×12 cm = 60 cm.

7. (D) 7
   3 marbles are white since $\frac{1}{8} \times 24 = 3$. There are $24 - 3 = 21$ non-white marbles. 14 of them are red since $\frac{2}{3} \times 21 = 14$. The other marbles are brown, so there are $21 - 14 = 7$ brown marbles.

8. (A) 2
   One move is not enough, since 3 numbers of the top row are not where we want them, and one move can change the positions of only 2 numbers. In two moves we can first switch 2 and 5, and then switch 2 and 3. At that point, we have the cards arranged in the way shown in the bottom row.

SOLUTIONS 2005

9. (E) 103
Tom's result is a multiple of 3. An easy way to see if a number is a multiple of 3 is to apply the rule that if a number is a multiple of 3, the sum of its digits is a multiple of 3.
   (A) $9 + 8 + 7 = 24$, which is a multiple of 3 ($987 = 329 \times 3$)
   (B) $4 + 4 + 4 = 12$, which is a multiple of 3 ($444 = 148 \times 3$)
   (C) $2 + 0 + 4 = 6$, which is a multiple of 3 ($204 = 68 \times 3$)
   (D) $1 + 0 + 5 = 6$, which is a multiple of 3 ($105 = 35 \times 3$)
   (E) $1 + 0 + 3 = 4$, which is not a multiple of 3
103 is the only number listed that cannot be the result of multiplication by 3.

10. (C) 1
A quarter of 24 hours is 6 hours, a third part of 6 hours is 2 hours, and half of 2 hours is 1 hour.

11. (C) 37
9 out of the first 10 pieces are put aside and the 10th piece is used for the next step. Eva cuts it into 10 new pieces and puts 9 of them aside. At this moment $(9 + 9)$ pieces are put aside and the 19th piece is used for the next step. Eva cuts it into 10 newer pieces and puts 9 of them aside increasing the total number of pieces put aside to $[(9 + 9) + 9]$. She finally cuts the 28th piece into 10 pieces. Altogether, Eva cut the napkin into $[(9 + 9) + 9] + 10 = 37$ pieces.

12. (D) 48 minutes
Mowgli needs 16 minutes to travel on the elephant in one direction since it took him 32 minutes to travel on the elephant in both directions. He needs (40 – 16) minutes = 24 minutes to travel in one direction on foot since it took him 40 minutes to travel in one direction on foot and in the other direction on the elephant. Hence, Mowgli needs 2×24 minutes = 48 minutes to travel both ways on foot.

13. (C) 8 m²

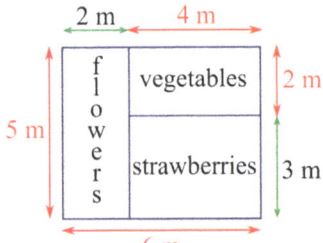

The section with flowers has an area of 10 m², so the side shown as vertical is $10 \div 2 = 5$ m long. This is also the width of the whole garden. The whole garden has an area of 30 m², so the garden's length is $30 \div 5 = 6$ m. The vegetable section is 6 m – 2 m = 4 m long. The length of the other side is 5 m – 3 m = 2 m. Therefore, the vegetable section has an area of 2 m × 4 m = 8 m².

14. (E) 7
Grandpa's distribution of peanuts can be written as 1×5 kg, 2×4 kg, 4×2 kg, 2×1.5 kg, and 1×0 kg. The number of family members is $1 + 2 + 4 + 2 + 1 = 10$ and the amount of peanuts to be distributed among them is
$(1 \times 5 + 2 \times 4 + 4 \times 2 + 2 \times 1.5 + 1 \times 0) = 5 + 8 + 8 + 3 + 0 = 24$ kg.
If 24 kg of peanuts were distributed equally among the 10 family members as Grandma suggested, then each would get 2.4 kg of peanuts. Grandma's suggestions would give 7 family members more peanuts than what Grandpa suggested.

SOLUTIONS 2005

15. (B) 20
13, 15, 17, 19; 31, 35, 37, 39; 51, 53, 57, 59; 71, 73, 75, 79; and 91, 93, 95, 97 are all 2-digit numbers that use only odd digits that are different. There are $5 \times 4 = 20$ of them.

16. (B)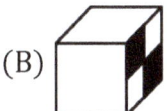

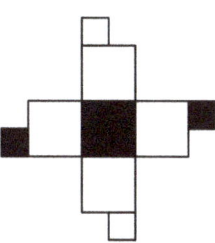

When you look at the net and keep the completely black face as the bottom of the folded cube, then all four side faces are completely white and the top face is checkered (opposite corners of the face have the same color). The completely black and checkered faces are opposite each other and no three completely white faces form a corner of the cube. These observations eliminate options (A), (C), (D), and (E). Only (B) is consistent with all of them.

17. (B) 403
Let m be the middle number. The five consecutive numbers are $m-2, m-1, m, m+1, m+2$. Their sum is $5m$, and $5m = 2005$. Hence, $m = 401$, and $m + 2 = 403$ is the greatest number.

18. (E) 9
$1 \times 100, 2 \times 50, 4 \times 25, 5 \times 20$, and 10×10 are each equal to 100, so 1, 100, 2, 50, 4, 25, 5, 20, and 10 are all different divisors of 100. There are 9 divisors of the number 100.

19. (C) 1 decimeter

The corners of the frame (shown in red) are squares with the length of each side equal to the width of the frame. Two outer sides of each of the 4 squares are segments of the outer perimeter. The other 4 segments of the outer perimeter are matched exactly by the corresponding 4 sides of the inner perimeter. The difference between the outer and inner perimeters consists of the two outer sides of 4 corner squares. 4×2 outer sides = $8 \times$ one outer side, which is $8 \times$ the width of the frame. That difference is 8 decimeters, so the width of the frame is 1 decimeter.

20. (E) 3 more
There are 6 squares of size 1×1 and 1 square of size 2×2, so there are 7 squares. The big triangle is a right triangle and its hypotenuse consists of 4 equal segments. All the other triangles are also right triangles with each hypotenuse being a segment of the largest triangle's hypotenuse. There are 4 triangles with a one-unit hypotenuse, 3 triangles with a two-unit hypotenuse, and 2 triangles with a three-unit hypotenuse. There are $1 + 4 + 3 + 2 = 10$ triangles, so there are $10 - 7 = 3$ more triangles than squares.

SOLUTIONS 2005

21. (E) 8
First, unlock the treasure chest and one container. If we unlock all three boxes in the unlocked container, we get only 30 golden coins. We need 50 golden coins, so we have to unlock another container and two boxes inside this container to get 20 more golden coins. At this point $1 + 1 + 3 + 1 + 2 = 8$ locks were opened and we have 50 golden coins.

22. (D) 82
To the left of 6 we have to insert 9 since $9 + 6 = 15$. To the left of 15 we have to insert 12 since $12 + 15 = 27$. To the left of the 9 we have to insert 3 since $3 + 9 = 12$. After this, we can use addition to fill in the rest of the circles. $5 + 3 = 8$; $7 + 8 = 15$; $8 + 12 = 20$; $15 + 20 = 35$; $20 + 27 = 47$; and $35 + 47 = 82$. So, 82 should replace x.

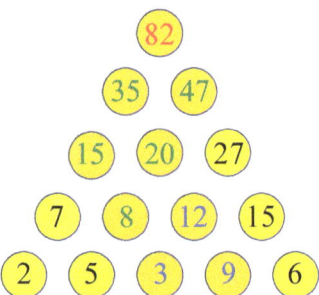

23. (D) $b = 2a$
A multiple of 6 is an even multiple of 3, so the last digit (the ones digit b) must be even **and** the sum of digits, $a + b$, must be a multiple of 3. If $b = 2a$, then b is even and $a + b = 3a$ is a multiple of 3. We can find a counterexample for each of the other conditions:
For (A) $a + b = 6$, $1 + 5 = 6$ but 15 is not a multiple of 6.
For (B) $b = 6a$, $6 = 6 \times 1$ but 16 is not a multiple of 6.
For (C) $b = 5a$, $5 = 5 \times 1$ but 15 is not a multiple of 6.
For (E) $a = 2b$, $2 = 2 \times 1$ but 21 is not a multiple of 6.
Thus, only $b = 2a$ ensures the two-digit number is a multiple of 6.

24. (D) 0.5 gallon

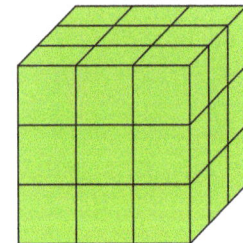

A $3 \times 3 \times 3$ cube consists of $3 \times 3 \times 3 = 27$ unit cubes. The sides of these unit cubes are small unit squares. There are 6×27 of them since every cube has 6 faces. The surface of the big cube consists of 6×9 of the small unit squares that were already painted with 0.25 gallon of paint, so $(6 \times 27 - 6 \times 9) = 6 \times 18 = 2 \times (6 \times 9)$ small unit squares are not painted yet. 0.25 gallon of paint is needed for 6×9 unit squares, so 0.5 gallon of paint is needed to paint the unpainted sides of the little cubes.

25. (D) $2 : 3$
The shaded area consists of one full circle and 4 quadrants, so the shaded area is equivalent to 8 quadrants. At each of the 4 vertices of the square there are 3 quadrants of the circle which are unshaded, so 12 quadrants are unshaded. $8 : 12$ equals $2 : 3$, so the ratio of the shaded area to the unshaded area is $2 : 3$.

26. (C) 18
The note is true from 2 p.m. to 11 p.m. and from 2 a.m. to 11 a.m., so it is true for 18 hours.

© Math Kangaroo in USA, NFP   www.mathkangaroo.org

SOLUTIONS 2005

27. (D) 3 cm

If a and b are the dimensions [in cm] of the base and h is the height of the prism, then half of the perimeter of the base is $a + b$, so $a + b = 9$. The volume of the prism is $(a \times b) \times h$, so $(a \times b) \times h = 42$. a, b, and h are natural numbers, $a + b = 9$ and $a \times b$ is a divisor of 42. $1 + 8 = 9$ but $1 \times 8 = 8$ is not a divisor of 42. $2 + 7 = 9$ and $2 \times 7 = 14$ **is** a divisor of 42. $3 + 6 = 9$ but $3 \times 6 = 18$ is not a divisor of 42. $4 + 5 = 9$ but $4 \times 5 = 20$ is not a divisor of 42. The only valid dimensions of the base are 2×7 (or 7×2), so the height is $42 \div 14 = 3$ cm.

28. (D) 4

Let's label the 3-digit number. The first digit is a, the second digit b, and the third digit c. We are told that the square root of the first digit is equal to the second digit divided by the third. We can write the equation $\sqrt{a} = b/c$.

Since we know that each of these numbers is a single digit, there are only several options for what the value of a could be. The only 1-digit numbers that are perfect squares are 1, 4, and 9; we can immediately rule out 1 since all the digits would have to be the same to meet the other condition.

Let's first look at 4 as the first digit. The square root is equal to 2. We must find the digits b/c that will give the value of 2. The values of b and c are 2 and 1 as well as 6 and 3. This way we know that Peter for sure wrote the numbers 421 and 463.

Next, we can look at the other digit: 9. The square root of 9 is 3. The only way that we can get b/c without repeating any digits is with $3 \div 1$ and $6 \div 2$. Therefore, we know that Peter wrote down the following numbers: 931 and 962.

This gives us a total of 4 numbers that Peter wrote: 421, 463, 931, and 962.

29. (B) 22.5

In each of the four consecutive rows of the shaded area (starting from the bottom) there are 4 full triangles and 2 halves of triangles, which is equivalent to 5 triangles. Thus, the shaded area in these 4 rows is equivalent to $4 \times 5 = 20$ triangles.

The next row consists of the parallelogram (shown in red) and one triangle to its right. Half of the parallelogram and half of the triangle are shaded, so half of the fifth row is shaded. The whole fifth row consists of 5 triangles, so $5 \div 2 = 2.5$ are shaded. Therefore, $20 + 2.5 = 22.5$ triangles are shaded. Each triangle has the area equal to 1, so the area of the shaded region is equal to 22.5.

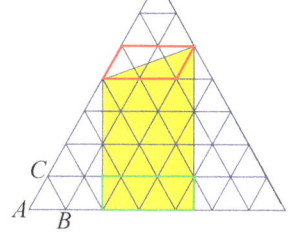

30. **(D) Ella**

Dorothy and Sylvia are not sitting the farthest to the left. Mary and Sylvia are not sitting the farthest to the right. Also, Dorothy is not sitting the farthest to the right since Ella is sitting to Dorothy's right. Thus, Dorothy and Sylvia are sitting in the middle 3 seats of the bench but Sylvia is not sitting next to Dorothy. Hence, neither Sylvia nor Dorothy is sitting at the very center of the bench.

One option is:

?	Sylvia	?	Dorothy	Ella

If this were the case, Kathy would be either farthest to the left or in the center. Either way, she would be sitting next to Sylvia, which is not the case. Therefore, this is not the arrangement.

Thus, the seats for Dorothy and Sylvia must be:

?	Dorothy	?	Sylvia	?

Kathy cannot sit next to Sylvia, so Kathy is sitting farthest to the left.

Kathy	Dorothy	?	Sylvia	?

Mary cannot be farthest to the right, so she must be in the center and Ella must be farthest to the right. The only way the girls can be sitting on the bench is shown below.

Kathy	Dorothy	Mary	Sylvia	Ella

Solutions for Year 2007

1. **(D) 223**

   Remember to do operations in parentheses first. Do multiplication and division before addition and subtraction.
   $2007 \div (2+0+0+7) - 2 \times 0 \times 0 \times 7 = 2007 \div 9 - 0 = 223 - 0 = 223$.

2. **(D) $\frac{3}{8}$**

   $1\frac{1}{2} = \frac{3}{2}$ and $\frac{3}{2} \times \frac{1}{4} = \frac{3}{8}$, so Donald spends $\frac{3}{8}$ of the day sleeping.

3. **(D) 120°**

   The sum of the measures of the three angles of any triangle is 180°. So, the measure of the third angle of the triangle is $180° - (12° + 48°) = 180° - 60° = 120°$.

4. **(C) 15 seconds**

   It takes the kangaroo 6 seconds to make 4 jumps, so in 12 seconds the kangaroo can make 8 jumps and in 3 seconds it can make 2 jumps. In 15 seconds the kangaroo can make 8 jumps + 2 jumps, which is 10 jumps.

SOLUTIONS 2007

5. (B) 6
Between 16:00 and 17:00, the sum of the digits of the hour part of the time is 7. The sum of the digits of the minute parts is also 7 at 16:07, 16:16, 16:25, 16:34, 16:43, and 16:52. Thus, it happens 6 times.

6. (B)
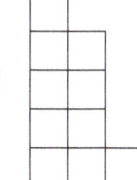

The solution is shown to the right. The original given piece is shaded, and the non-shaded region is piece (B) rotated 90° counterclockwise.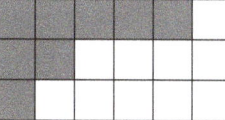

(A) 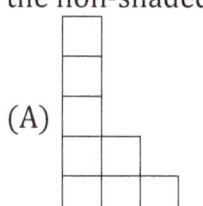   (B) 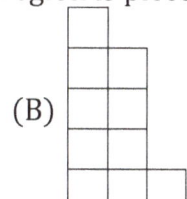   (C) 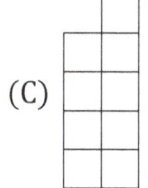   (D) 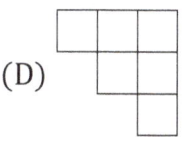   (E)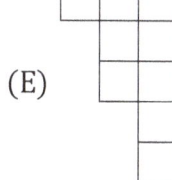

The rectangles that can be formed must have a height of at least 3 and a width of at least 5. Any such rectangle has an area of at least 15. This eliminates (D) since the shaded piece and piece (D) cover only the area of $8 + 6 = 14$ unit squares.
Piece (A) and the shaded piece together contain $8 + 8 = 16$ unit squares. Possible dimensions of rectangles with the area of 16 are 1×16, 2×8, and 4×4. However, each of the dimensions for the rectangle we are forming must be at least 3 and at least one of them must be at least 5. Thus, piece (A) and the shaded piece cannot form a rectangle.
Both piece (C) and piece (E) contain 9 unit squares each. $8 + 9 = 17$, so 1×17 are the only possible dimensions for any rectangle with the area of 17. So, neither piece (C) nor piece (E) fit next to the shaded piece.

7. (A) 1
Each step below is unique, so there is only one solution to this puzzle.

8. (D) December 31st, 2002
Ian is older than Peter, so Peter was born later. Ian was born on January 1st, 2002, so exactly one year later it was January 1st, 2003. 1 day prior to January 1st, 2003, was December 31st, 2002, so this is the date of Peter's birth.

© Math Kangaroo in USA, NFP   93   www.mathkangaroo.org

SOLUTIONS 2007

9. (C) 16
The sum of the ages of all three girls is 32 but Alexa is 8 years older than each of the twins. To compensate for the age difference, we add 8 + 8 to 32, which gives us the triple of Alexa's age. 8 + 8 + 32 = 48, so Alexa's age is 48 ÷ 3 = 16. Alexa is 16 years old and each twin sister is 8 years old.

10. (A) 100 m
1 meter = 10 decimeters (1 m = 10 dm), so a given cube is cut into 10 × 10 × 10 = 1000 smaller cubes. The height of each smaller cube is 1 dm, so the height of the structure of 1000 of them, stacked one on top of another, is 1000 × 1 dm = 100 × 10 dm = 100 × 1 m = 100 m.

11. (D) 14 inches
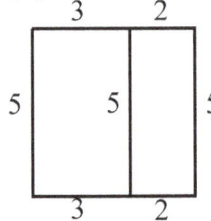
Each side of a square with a perimeter of 20 inches has a length equal to 5 inches because 5 × 4 = 20. A rectangle with two parallel sides equal to 5 inches each and with a perimeter equal to 16 inches has the other two sides equal to 3 inches each since 5 + 3 + 5 + 3 = 16. This means that the other rectangle has shorter sides each equal to 5 - 3 = 2 inches. It has a perimeter equal to 14 inches since 5 + 2 + 5 + 2 = 14.

12. (C) 25

    The number of colored small squares along each diagonal corresponds to the number of rows (or columns) in the larger square grid.

If the dimensions of the grid are even (2 × 2, 4 × 4, 6 × 6, 8 × 8, ...), the number of all the colored unit squares is exactly the number of columns + the number of rows (4, 8, 12, 16, ... and so on). If the dimensions of the grid are odd (1 × 1, 3 × 3, 5 × 5, 7 × 7, ...), the unit square in the center is counted twice, so we have to subtract 1 from the number of columns + the number of rows. In this case the number of colored unit squares is 1, 5, 9, 13 ... and so on. 9 unit squares are colored when there are 5 rows and 5 columns, because 5 + 5 − 1 = 9. Then, there are 5 × 5 = 25 unit squares making up the grid.

13. (C) Carl plays volleyball.
Alex plays neither soccer nor volleyball and Ben does judo, so Alex does karate. Hence, "Alex plays volleyball," "Alex does judo," "Ben plays soccer," and "Daniel does karate" are false statements. The only statement left is "Carl plays volleyball," so it is the only statement that could be true.

SOLUTIONS 2007

14. (B) 54

The visible edges of the rectangular section of the block are shown in red. The top face of the section is removed from the surface of the block but its area is matched by the area of bottom face of the section. The left face of the section is removed from the surface of the block but its area is matched by the area of the right face of the section. The front and back faces of the section are also removed from the surface of the block and, and since there is no compensating for them, this will be the difference between the surface area of the block before and after the removal of the rectangular section.

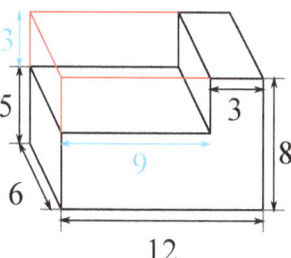

The width of the block is 12 and 12 − 3 = 9, so the width of the section is 9.
The height of the block is 8 and 8 − 5 = 3, so the height of the section is 3.
Hence, 3 × 9 is the area of the front face of the section. The same is true for the back face, so the surface area of the block decreases by 2 × (3 × 9) = 54.

15. (B) 13.5 inches

Within each rectangle the red segment and the black segment are equal since the red dot is the center of the rectangle. Hence, the combined length of the black line segments and the combined length of the red line segments are equal. Therefore, the sum of the lengths of the original line segments is half of 27 inches, which is 13.5 inches.

16. (C) 22

The number of birds that flew away from the three trees is 6 + 8 + 4. Thus, 60 − (6 + 8 + 4) = = 60 − 18 = 42 birds stayed in these 3 trees. The same number of birds was left on each tree, so 42 ÷ 3 = 14 birds were left on each tree. 8 birds flew away from the second tree, so at the beginning 14 + 8 = 22 birds were on the second tree.

17. (B) 45 in²

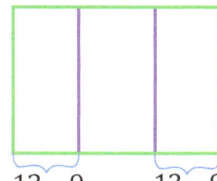

The region where the two 9 × 9 squares overlap is a rectangle between the two purple segments. Its height is 9 inches. The area of the overlap in square inches is:

$$9 \times \left(13 - \left((13-9) + (13-9)\right)\right) = 9 \times (13 - (4+4)) = 9 \times 5 = 45$$

18. (C) 40

The difference between 7:30 a.m. and 9:10 a.m. is 30 minutes + 60 minutes + 10 minutes, which is 100 minutes or 10 × 10 minutes. The pigeon flies 4 miles in 10 minutes, so it flew 40 miles during its 100-minute flight from 7:30 a.m. to 9:10 a.m.

© Math Kangaroo in USA, NFP   www.mathkangaroo.org

SOLUTIONS 2007

19. (D) D1

All the turns are 90° right turns until the kangaroo reaches the square D1 and cannot continue moving forward even after making a right turn.

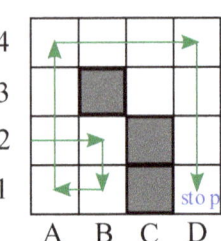

20. (B) 50

Right now, Anna's mother is 40 years old since $4 \times 10 = 40$. Anna will be 20 years old when she is twice as old as she is right now, and this will happen in 10 years. In 10 years, Anna's mother will be 50 years old since $40 + 10 = 50$.

21. (D)

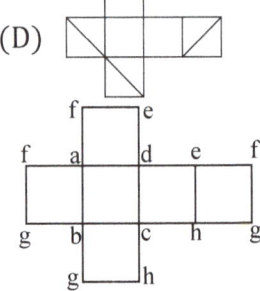

For each grid we list endpoints of its unit square diagonals according to the labeling of the vertices shown to the left. The same label indicates that the vertices become one vertex of the cube when the grid is folded.

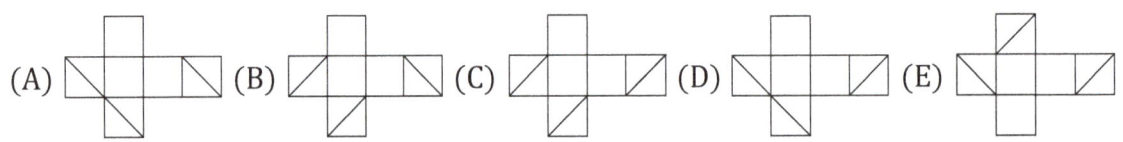

Without repetition, the endpoints of the diagonals are:

(A) f, b, h, e, g   (B) a, g, c, e   (C) a, g, c, f, h   (D) f, b, h   (E) f, b, e, a, h

Only (D) is restricted to three vertices, so the other options are not valid representations of the diagonals of cube faces (shown below to the left with labels).

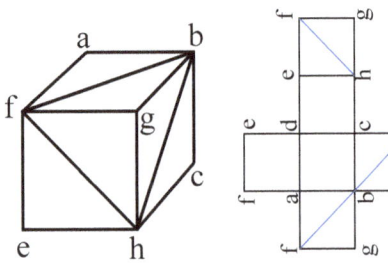

The diagram (D) with labels is shown after 90° rotation for easier comparison. The face h-e-f-g of (D) becomes the front face of the cube, and the face h-e-d-c of (D) is pushed back and becomes the bottom face of the cube.

Then the face c-d-a-b moves up and becomes the back face of the cube, the face b-a-f-g comes on the top of the cube, and the other two faces become the right and left faces of the cube.

© Math Kangaroo in USA, NFP

SOLUTIONS 2007

22. **(B) 101**
The question is asking about creating 4-digit numbers by rewriting a 2-digit number next to itself. We can rewrite any four-digit number based on the place value of each digit: for example, 4375 is equal to $4 \times 1000 + 3 \times 100 + 7 \times 10 + 5 \times 1$. If we begin with a 2-digit number and call the digits *ab*, when we rewrite the digits to create a 4-digit number we get *abab*. We know that *abab* is equal to $a \times 1000 + b \times 100 + a \times 10 + b \times 1$ or $1010a + 101b$, which will simplify further into $101(10a + b)$. Looking at the value inside the parentheses, $10a + b$, it can be observed that any 2-digit number *ab* can be rewritten as $10a + b$, so the value inside the parenthesis represents any 2-digit number. The 101 which we multiply by $10a + b$ represents how many times the new four-digit number is greater than the original two-digit number.

23. **(D) P_2 and P_1 have equal perimeters.**
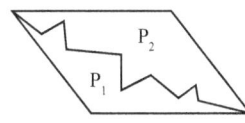
In any parallelogram the opposite sides are of equal length, and the segments within the parallelogram are sides of both parts, so P_2 and P_1 have equal perimeters. Visually, P_2 has a larger area than P_1, and there is no rule stating whether the areas will be equal or which will be larger.
Hence, only option (D) is a true statement.

24. **(E) O**
KANGAROO consists of 8 letters, so each letter repeats every 8 steps. $2000 = 250 \times 8$, so after the 2000th position the letters of the sequence are KANGARO.... The 2007th letter is O.

25. **(E) 50 inches**
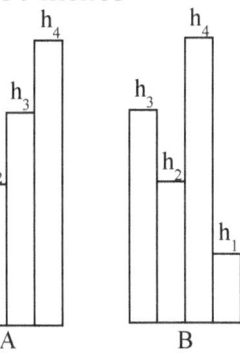
Figure A and figure B have the same number of horizontal segments, so we can ignore these segments since they do not influence the difference between the perimeters. For figure A the sum of vertical distances on the left side is exactly the same as the vertical distance on the right side, which is h_4.
Thus, the vertical distance of figure A is $2h_4$.
For figure B the sum of vertical distances is
$h_3 + (h_3 - h_2) + (h_4 - h_2) + h_4$ since $(h_4 - h_1) + h_1$ is equal to h_4.
$h_3 + (h_3 - h_2) + (h_4 - h_2) + h_4$ simplifies to $2h_3 - 2h_2 + 2h_4$.
The difference between the perimeters is $[2h_3 - 2h_2 + 2h_4] - 2h_4 =$
$= 2h_3 - 2h_2 = 2 \times (h_3 - h_2)$, which is 50 inches since $h_3 - h_2 = 25$ inches.

26. **(B) 72 inches**
Each square has one bold side and three thin sides of the same length, so the ratio is 1 to 3 between the bold sides adding up to 24 inches and the thin line $AA_1A_2A_3 ... A_{10}A_{11}A_{12}B$. So, the length of the thin line is 3×24 inches $= 72$ inches.
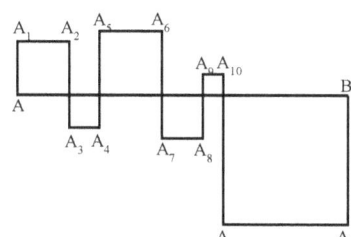

SOLUTIONS 2007

27. (C) 12

Let t be Tom's integer. Robert's possible number is 5t or 6t.
Dan's possible number is 5t + 5, 6t + 5, 5t + 6, or 6t + 6. Adam's possible number is
5t + 5 − 5, 6t + 5 − 5, 5t + 6 − 5, 6t + 6 − 5, 5t + 5 − 6, 6t + 5 − 6, 5t + 6 − 6, or 6t + 6 − 6. After simplification and removing duplications, Adam's possible number is 5t, 6t, 5t + 1, 6t + 1, 5t − 1, or 6t − 1. Adam's actual number is neither 5t nor 6t since 73 is neither a multiple of 5 nor 6. Adam's actual number is neither 5t − 1 nor 6t − 1 since 5t − 1 = 73 and 6t − 1 = 73 are equivalent to 5t = 74 and 6t = 74, and 74 is neither a multiple of 5 nor 6. Also, 5t + 1 cannot be 73 since 5t + 1 = 73 is equivalent to 5t = 72 and 72 is not a multiple of 5.
The last possibility is 6t + 1 = 73 and its solution is t = 72 ÷ 6 = 12. Thus, Tom's integer is 12.

28. (D) 10 inches

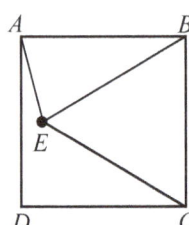

∠EAB = 75° and ∠ABE = 30°, so ∠BEA = 180° − (75° + 30°) = 75°. Hence, the triangle ABE is isosceles with BE = BA. BA and BC are sides of a square, so BA = BC and, consequently, BE = BC. BE = BC, so ∠BCE = ∠BEC. ∠EBC = 90° − ∠ABE = 90° − 30°, which is 60°. The other two angles of the triangle EBC are equal, so all three angles are equal to 60°. Thus, the triangle EBC is an equilateral triangle, and so EC = EB = BC = 10 inches.

29. (A) 1

We want to show that triangles ADE and BCF are equal, and that triangles ABE and DCH are equal. Consequently, the white area of the square ABCD is equal to the shaded area outside the square ABCD. Therefore, the area of the whole square ABCD is equal to the whole shaded region, which is 1.
Remember that *if two opposite sides of a quadrilateral are equal and parallel, then the other two opposite sides are also equal and parallel, and so the quadrilateral is a parallelogram.*

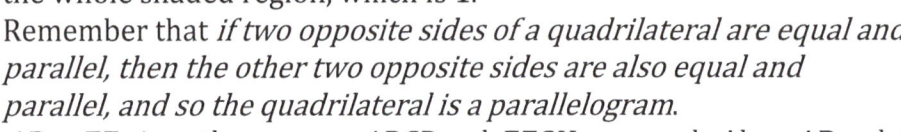

AB = EF since the squares ABCD and EFGH are equal. Also, AB and EF are parallel.
Hence, in the quadrilateral ABFE, AE = BF and are parallel.

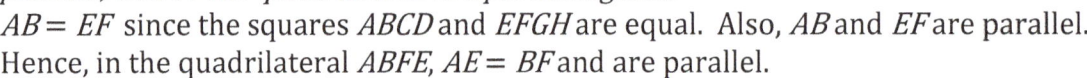

DC is parallel to AB (opposite sides of the square ABCD) and AB is parallel to EF, so EF and DC are parallel. DC = AB (opposite sides of ABCD) and AB = EF, so DC = EF and are parallel.
Hence, DEFC is a parallelogram and DE = CF.
Consequently, triangles ADE and BCF are equal since AD = BC, DE = CF, and EA = FB.
In the quadrilateral DAEH, DA = HE and are parallel, so EA = HD. Of course, AB = DC.
In the quadrilateral HEBC, HE = CB and are parallel, so BE = CH.
Therefore, triangles ABE and DCH are equal since AB = DC, BE = CH, and EA = HD.
In summary: Triangles ADE and BCF are equal, the triangles ABE and DCH are equal, and the area of ABCD is 1.

SOLUTIONS 2007

30. (A) 5

The die with the question mark on its top face has 6 dots on its front left face since this front face touches the back right face of the die with 1 dot on its front face and $1 + 6 = 7$. The front right face of the die with the question mark has 3 dots since it touches the face opposite the face with 4 dots visible.

Look at the die with the question mark. Put your right hand along its left front face (the one with 6 dots) in such a way that your curling fingers go along the face with 3 dots, then your thumb points toward the face with the question mark. Apply the same right-hand rule, $6 \Rightarrow 3 \Rightarrow$ another face, to the identical die with the three visible faces. Your thumb is pointing toward the right back face. This face has 5 dots since the left front face has 2 dots and $5 + 2 = 7$. Thus, $6 \Rightarrow 3 \Rightarrow 5$ is the outcome of the right-hand rule. All the dice are identical, so there are 5 dots on the face with the question mark.

Solutions for Year 2009

1. (D) 200×9

2009 and 9 are odd numbers and 0 and 200 are even numbers. *Even ± odd* is always an odd number. Any number multiplied by an even number always yields an even number. Thus, $2 + 0 + 0 + 9$, $200 - 9$, and $200 + 9$ are odd numbers. Only 200×9 is an even number.

2. (B) In the circle and in the square, but not in the triangle.

3. (B) 17

The integers between 2.009 and 19.03 are 3, 4, ..., 18, and 19. There are $19 - 2 = 17$ of them.

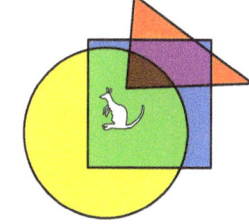

4. (C) 3

12321 is an example of a palindrome (a number or word that reads the same backward and forward) obtained from 12323314 by erasing 33 and 4. Another example of a 5-digit palindrome is 13331.

Since there is only one 4, any palindrome containing 4 would need to have 4 at its center, so 4 must be erased. We are left with seven digits, 1232331. This is not a palindrome. If you erase just one more digit from 1232331 none of the outcomes (232331, 132331, 122331, 123331, 123231, 123231, or 123233) end up being palindromes. There are no palindromes with more than 5 digits, so 3 is the smallest number of digits that need to be erased.

5. (A) white

The apple is neither in the white nor in the green box, so it must be in the red box.
The chocolate bar is either in the white or in the red box, but we already know that the red box contains the apple, so the chocolate bar must be in the white box.

© Math Kangaroo in USA, NFP   99   www.mathkangaroo.org

SOLUTIONS 2009

6. (B) 105°
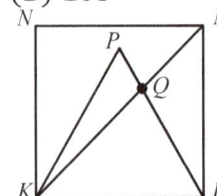
∠KLQ = ∠KLP = 60° since the triangle KLP is equilateral.
∠QLM = 90° − 60° = 30°.
∠QML = ∠KML = 45° since KM is a diagonal of the square KLMN.
∠LQM = 180° − (∠QML + ∠QLM) = 180° − (45° + 30°) = 105°.
Therefore, the measure of ∠LQM is 105°.

7. (D) 240 meters
$\frac{1}{4} + \frac{1}{4} = \frac{1}{2}$ of the bridge is over land, so $\frac{1}{2}$ of the bridge is over the river.
The river is 120 meters wide, so the bridge is 2 × 120 meters = 240 meters long.

8. (C) 420 cm

When moving along the bold line path, starting at the top side of the rectangle, we move through s-s-m-m-m-m-m-ℓ-ℓ-s-s-s segments, where s stands for short (s = 20 cm), m for medium (twice the length of s), and ℓ for long (three times the length of s). The length of the whole path is:
5s + 5m + 2ℓ = 5s + 5 × (2s) + 2 × (3s) = 21 × s = 21 × 20 cm = 420 cm

9. (C) half the number of dogs
A cat has 4 paws and a dog has one nose, so 4 × the number of cats = 2 × the number of dogs in the room. The number of dogs has to be twice the number of cats to make the two sides equal. Hence, the number of cats is half the number of dogs.

10. (E) 14
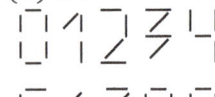
Count the number of sticks needed to make each digit. The digit 8 requires 7 sticks, which is more than the number of sticks needed for any other digit. Therefore, 88 requires the most sticks, which is 14. Thus, the greatest number written by Stan is 14.

11. (C) 6
78 = 2 × 3 × 13, so the divisors of 78 are: 1, 2, 3, 6, 13, 26, 39, and 78. n is positive, so n + 2 is a divisor of 78 that is greater than 2. Subtract 2 from the divisors greater than 2 to get n. The corresponding values for n are: 1, 4, 11, 24, 37, and 76. Thus, there are 6 positive whole numbers n such that n + 2 is a divisor of 78.

12. (C) 48
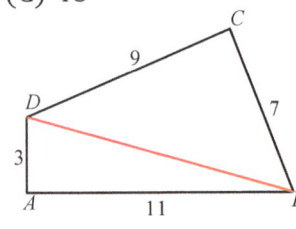
We can divide the quadrilateral into two right triangles.
|DA| = 3, |AB| = 11, and ∠DAB = 90°, so the area of the triangle DAB is ½ × 3 × 11. |CD| = 9, |BC| = 7, and ∠DCB = 90°, so the area of the triangle DCB is ½ × 9 × 7. The sum ½ × 3 × 11 + ½ × 9 × 7 = = ½ × (33 + 63) = ½ × 96 = 48 is the area of the quadrilateral ABCD.

SOLUTIONS 2009

13. (D) 174
Each week the difference between the number of boys and the number of girls decreased by 2 since 8 − 6 = 2. The original difference was 39 − 23 = 16, so after 8 weeks (8 × 2 = 16) the difference was zero and the number of girls was the same as the number of boys. At that time, there were 23 + 8 × 8 = 87 girls and 39 + 8 × 6 = 87 boys. Therefore, there were 174 participants in the dance group at that time.

14. (E) 65

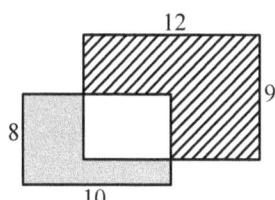

The area of the overlap is the area of the smaller rectangle minus the shaded area, so it is (8 × 10) − 37 = 80 − 37 = 43. The area of the region covered with diagonal lines is the area of the bigger rectangle minus the area of the overlap, so it is (9 × 12) − 43 = 108 − 43 = 65. Notice that the area of the region covered with diagonal lines can be also computed directly as (9 × 12) − (8 × 10) + 37.

15. (D) the card with 2 is in group *B*.
The sum of all numbers from 1 to 8 is 36 ([1 + 8] + [2 + 7] + [3 + 6] + [4 + 5] = 4 × 9), so the sum of numbers in group *A* is 36 ÷ 2 = 18. If 2 is in *A*, then the sum of the other two numbers in group *A* must be 16. If there are only three cards in group *A*, this is impossible since the largest possible sum of numbers on two cards is 8 + 7 = 15, which is less than 16. Therefore, we know for sure that the card with 2 is in group *B*.
Possible sums of numbers in *B* are: 1 + 2 + 3 + 4 + 8, 1 + 2 + 3 + 5 + 7, and 1 + 2 + 4 + 5 + 6. This means that all the other options—(A), (B), (C), and (E) —for group *B* can be false.

16. (C) 6 cm
The perimeter of the square is 4 × 9 cm = 36 cm. The perimeter of the equilateral triangle is the same, so one side of the triangle has the length of 36 cm ÷ 3 = = 12 cm. Half of the perimeter of the rectangle is 18 cm and its longer side (shared with the triangle) has the length of 12 cm, so the length of the shorter side of the rectangle is 6 cm.

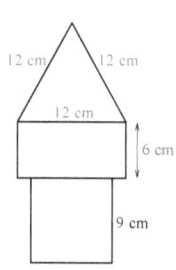

17. (B) 12
To use the smallest number of identical cubes to make a rectangular prism with the dimensions of 40 × 40 × 60 we need cubes that are as large as possible. The length of the edge of the cubes must be a divisor of 40 and a divisor of 60. The greatest common divisor of 40 and 60 is 20, so we use cubes with the dimensions of 20 × 20 × 20. We need 12 such cubes since 40 × 40 × 60 = 2 × 20 × 2 × 20 × 3 × 20 = 12 × (20 × 20 × 20).

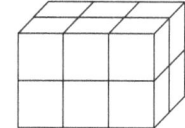

18. (E) 41
In one week, from Sunday to Saturday, Adam reads 25 + 6 × 4 = 49 pages of a 290-page book. During 5 full weeks he will read 5 × 49 = 245 pages, which is 45 pages short of the whole book. Then on Sunday he will read 25 pages. For the final 20 pages of the book he needs 5 more weekdays, so it will take him 5 × 7 + 1 + 5 = 41 days to read the whole book.

© Math Kangaroo in USA, NFP

SOLUTIONS 2009

19. (D) Daniel
$1 + 2 + 3 + 4 = 10$ and the sum of the numbers of the places of Adam, Bart, and Daniel is equal to 6, so Caesar took 4th place. The sum of the numbers of the places of Bart and Caesar is equal to 6, and we already know that Caesar placed 4th, so Bart took 2nd place. Adam did not win since Bart did better than he, so Adam took 3rd place and Daniel won 1st place.

20. (C) 3
$2009 = 7 \times 287 = 7 \times 7 \times 41$, which are the prime factors of 2009. Thus, the dimensions 1×2009, $7 \times (7 \times 41)$, and $41 \times (7 \times 7)$ are different and are all possible dimensions of the rectangles. The dimensions of the three different rectangles are: 1×2009, 7×287, and 41×49.

21. (C) 52
Jane's final result is $8^{18} \times 5^{50} = (2^3)^{18} \times 5^{50} = 2^{54} \times 5^{50} = 2^4 \times 2^{50} \times 5^{50} = 2^4 \times (2 \times 5)^{50} = = 16 \times 10^{50}$, which is 16 followed by fifty zeros. Thus, Jane's final product has 52 digits.

22. (B) 5
0 is divisible by 5, 11, 55, and also less than 10, so all four statements are true for 0. So, n cannot be (A) 0.
We have to find which two statements are true and which two are false.
If the 3rd statement (n is divisible by 55) is among the two true statements, then the 1st and 2nd statements (n is divisible by 5 and n is divisible by 11) are also true, so the first three statements would be true. Thus, n is divisible by 55 must be a false statement.
If the 1st and 2nd statements (n is divisible by 5 and n is divisible by 11) are true, then the 3rd statement is also true since $55 = 5 \times 11$ and the only common divisor of 5 and 11 is 1. Therefore, from the list of the first three statements only one can be true. It is not the third statement.
The other true statement must be the last one (n is less than 10).
Among the multiples of 5, 11, or 55, only 5 is less than 10, so n must be 5.

23. (C) 17
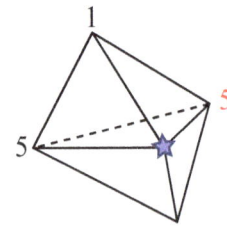
The edge connecting the vertex with 1 and the vertex with the blue star is the common edge of two triangular faces. The other two vertices of the two triangular faces are connected by the dashed edge.
When computing the sum of numbers at the vertices of the first face we are adding 1 + the number at the star + 5 and when computing the sum of numbers at the vertices of the second face we are adding 1 + the number at the star + the number at the other end of dashed edge. The sums are equal, so the numbers at the vertices of the dashed edge must both be 5. The third top face of the solid has 1, 5, and 5 assigned to its vertices. To keep the same sums for all the faces we have to assign 5 to the blue star and 1 to the fifth vertex of the solid. The sum of numbers at all five vertices is $1 + 5 + 5 + 5 + 1 = 17$.

SOLUTIONS 2009

24. (E) 105
2 as a 1st digit occurs only on the second floor and it occurs 35 times. 2 as a 2nd digit occurs on each of the five floors from 20 to 29, so 2 as a 2nd digit occurs 5 × 10 = 50 times. 2 as a 3rd digit occurs in _02, _12, _22, and _32 on every floor, so it occurs 5 × 4 = 20 times.
Therefore, the digit 2 was used 35 + 50 + 20 = 105 times.

25. (C) 48 cm²
The corner squares are identical, $|DC|$ = 10 cm and $|NM|$ = 6 cm, so $|DN| = |MC| = $ (10 cm – 6 cm) ÷ 2 cm = 2 cm. The area of the four corner squares is 4 × 2 cm × 2 cm = 16 cm².
The isosceles triangle with the base NM is a right triangle, so its vertical height (shown in red) is half of its base and the area of the triangle is $\frac{1}{2}$ × 6 cm × 3 cm = 9 cm². The area of four such triangles is 4 × 9 cm², which is 36 cm². Therefore, the area of the shaded region inside the square $ABCD$ is [100 – (16 + 36)] cm², which is 48 cm².

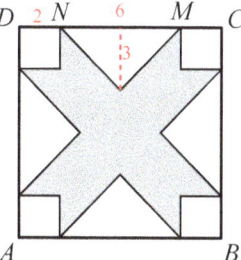

26. (C) 6
Subtract the second row from the first row:
$$(a + b + a) - (b + a + c) = 11 - 8$$
$$a - c = 3$$
According to the third column $2a + c = 9$. Add these two equations to get
$$(a - c) + (2a + c) = 3 + 9$$
$$3a = 12$$
$$a = 4$$
Since $a - c = 3$ then $c = 1$. From any row we can compute that $b = 3$. Therefore, $a + b - c = 4 + 3 - 1 = 6$.

27. (E) 168
There are 7 different numbers of dots (0, 1, 2, 3, 4, 5, and 6). A given number of dots occurs once with each of the 6 other numbers of dots and once as a double. Hence, in a complete set of dominoes, each number of dots occurs 8 times. Therefore, there are
8 × (0 + 1 + 2 + 3 + 4 + 5 + 6) = 8 × 21 = 168 dots altogether in a complete set of dominoes.

© Math Kangaroo in USA, NFP   www.mathkangaroo.org

SOLUTIONS 2009

28. **(D) 20**

Look at any row in a table with the left entry m and the right entry n. The second row contains $m + n$ and $m - n$ and the third row contains the sum and the difference of $m + n$ and $m - n$. Notice that $(m + n) + (m - n) = 2m$ and $(m + n) - (m - n) = 2n$, so the third row is double the entries of what is two rows above.

m	n
$m+n$	$m-n$
$2m$	$2n$

12	8
24	16
48	32
96	64

Thus, we double the entries of the table when we move down through every other row. We halve the entries of the table when we move up through every other row. If the numbers in 7th row are 96 and 64, the numbers in the 5th row are 48 and 32, the numbers in the 3rd row are 24 and 16, and the numbers in the 1st row are 12 and 8. So, the sum of numbers in the first row is $12 + 8 = 20$.

29. **(A) 5**

Each friend is labeled S-2 if the difference between their shoe sizes is 2 and is labeled S-1 if the difference between their shoe sizes is 1. Among all shoes bought by a group of friends, only the smallest right shoe and only the largest left shoe will be left over. All pairs of shoes bought by a group of friends have sizes between 36 and 45. 36 is an even number and 45 is an odd number, so we cannot go from size 36 to 45 by adding only 2 shoe sizes, which could happen if everybody was S-2. At least once we have to add 1 shoe size. This means that at least one person among the friends is S-1. For the smallest number of people there must be only one S-1 friend among the group. The options for the shoe sizes are: 36→37→39→41→43→45; 36→38→39→41→43→45; 36→38→40→41→43→45; 36→38→40→42→43→45; and 36→38→40→42→44→45.

Each case involves 6 pairs of shoes among 5 friends.

30. **(A) color A only**

We can start by putting either D or A in the empty square in the first row. Then, in the middle of the second row we can put either A or D, depending on how we started. In each case the table is completely determined by our choice as shown below in steps.

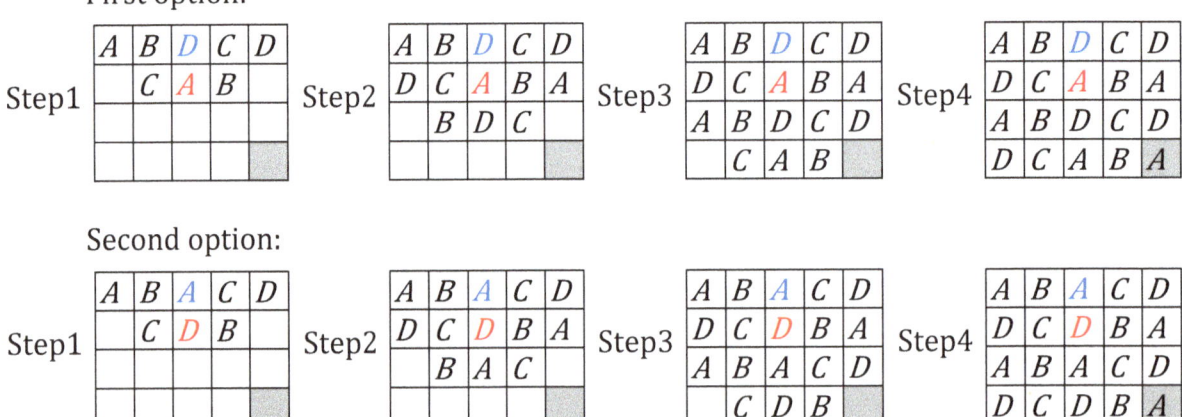

The shaded square can be colored as A only.

© Math Kangaroo in USA, NFP 104 www.mathkangaroo.org

Solutions for Year 2011

1. **(C) Wednesday**
    There are seven days in a week from Wednesday to Tuesday and eight letters in the word KANGAROO, so Basil will finish writing the word on the following Wednesday.

2. **(B) 56 km**
    Driving at the same speed and covering 28 km in half an hour, the motorcyclist will drive twice this distance, or 56 km, in one hour.

3. **(D) 36**
    A cube has six faces, so six cubes will have a total of $6 \times 6 = 36$ faces.

4. **(A) square**
    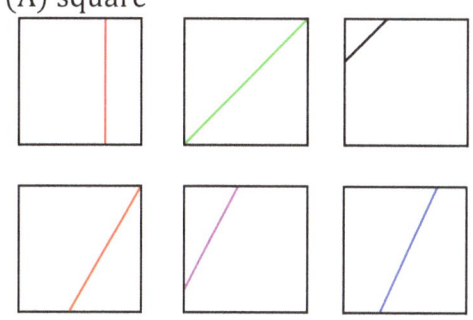
    The first picture shows a red line parallel to the sides of the square. If this is the line we cut along, both pieces are rectangles and neither of them is a square. If the cutting line is one of the diagonals, it divides the square into two right triangles that are isosceles (look at the green line). If the cutting line is parallel to one of the diagonals and divides the square into two pieces but is not one of the diagonals, then it divides the square into a right isosceles triangle and a pentagon (look at the black line). Any other line that divides the square into two pieces either passes through one of the vertices or cuts two sides of the square. In the first case one piece is a right triangle and the other piece is a right trapezoid (see the orange line). In the second case there are two options. If the line cuts two adjacent sides of the square, then one piece is a right triangle and the other piece is a pentagon (purple line). If the line cuts through parallel sides, then both pieces are right trapezoids (blue line). In summary, a square is never the result of any cut along a straight line that divides a square into two pieces.

5. **(E) 47**
    Let's list odd numbers and then cross out any that contain the digit 3. Here is the sequence that gives us the first 15 numbered houses on the right side of the street:
    1, ~~3~~, 5, 7, 9, 11, ~~13~~, 15, 17, 19, 21, ~~23~~, 25, 27, 29, ~~31, 33, 35, 37, 39~~, 41, ~~43~~, 45, |47|.

6. **(B) 750**
    The water flowing down will divide into equal amounts at each fork. 1000 gallons of water will first divide into 500 gallons flowing along each pipe, and then again divides in half at the next fork. 250 gallons will reach container X, and the rest (750 gallons) will pour into Y.
    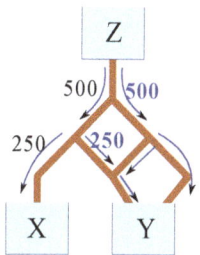

SOLUTIONS 2011

7. (B) 10 in²

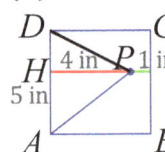

The base of triangle APD is |AD| = 5 in. Draw the height PH from point P to the triangle's base. The height |PH| = 4 in since P is one inch away from BC. The area of the triangle APD = ½ × (5 in) × (4 in) = 10 in².

8. (D) 15°

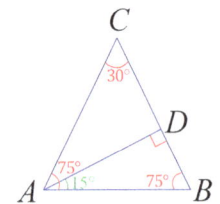

∠CAB = ∠CBA = (180° − 30°) ÷ 2 = 75° since |AC| = |BC| and ∠ACB = 30°.
AD is a height of the triangle, so ∠BDA is a right angle.
Now, ∠BAD + ∠CBA + ∠BDA = 180°, so ∠BAD = 180° − 90° − 75° = 15°.

9. (E)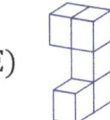

The big solid shown to the right is missing one cube in the top back row, three cubes in the top front row, and one cube from the row below. Stretch your right hand along the **three** missing cubes with the thumb pointing to the missing cube from the back row, then curl the fingers so they point toward the missing cube of the row below the top one. This right-hand rule identifies only piece (E) as the missing part of the prism. Notice that piece (E) can fit in two different ways to make the prism.
Piece (C) seems to be another possible option but it is a mirror image of piece (E), so it is not the missing piece.

10. (D) 1120 ml

On a day when Maria's cat catches a mouse, he drinks 60 ml + ⅓ of 60 ml of milk, which is 60 ml + 20 ml = 80 ml of milk. During the last 14 days (two weeks) the cat drank 14 × 80 ml = 1120 ml of milk.

11. (D)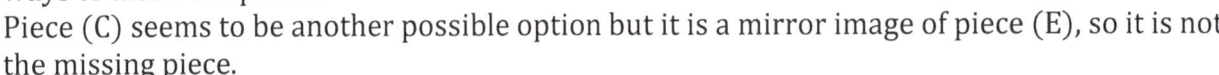

The pictures below show the order Andrew could have followed for answers (A), (B), (C), and (E). In answer (D), the cell with the letter G has no common point with the cell with the A that has not yet been marked, so Andrew cannot have written the word KANGAROO as shown in (D).

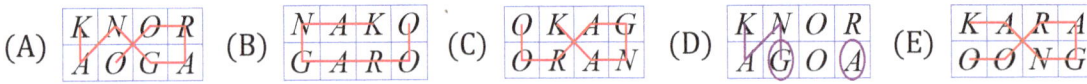

12. (B) 891

All four-digit numbers made by rearranging the digits of 2011 in increasing order are: 1012, 1021, 1102, 1120, 1201, 1210, 2011, 2101, 2110. The difference between the two numbers on either side of 2011 is 2101 − 1210 = 891.

SOLUTIONS 2011

13. (E) ⬚ 167

From numbers 17, 30, 49, 96, and 167, we have to select four numbers such that the sum of three of them equals to the fourth one. Of course, the sum is the largest of the four numbers. The sum is either 167 or 96, which are the two largest numbers. Actually, 96 = 17 + 30 + 49. To get 167 as the sum, we have to select 167, 96, and two other numbers. The sum of the two other numbers must be 167 − 96 = 71 but 17 + 30 = 47, 17 + 49 = 66, and 30 + 49 = 79. None of the last three results are 71, so there is no way to pick three numbers from the box to get the sum of 167. Therefore, 167 remains in the box.

14. (C) 64

4 cubes (marked in red) are used for the 4 corners, so 36 − 4 = 32 cubes match the 4 sides of the enclosed square region. Therefore, each side of the enclosed region is 32 ÷ 4 = 8 cubes in length on each side. 8 × 8 = 64 cubes are needed to fill in the interior of the square.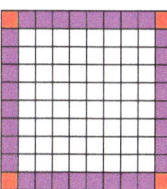

15. (B) 3,612

Paul's original number was 372 ÷ 31 = 12. If he had multiplied it by 301, his result would have been 12 × 301 = 3,612.

16. (D) 56 cm²

In all four right triangles the longer leg is 28 cm ÷ 2 = 14 cm and the shorter leg is 30 cm − (14 cm + 14 cm) = 2 cm. Therefore, the area of one triangle is ½ × 14 cm × 2 cm = 14 cm², and the sum of the areas of the four triangles is 4 × 14 cm² = 56 cm².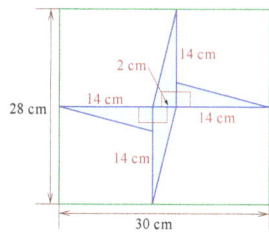

17. (B) 3 : 0

FC Barcelona played three games in which they scored a total of 3 goals and had 1 goal scored against them. The game which they lost had a result 0 : 1 (the only goal scored against FC Barcelona). The one game which was tied had to end 0 : 0, and therefore the game which FC Barcelona won had a score 3 : 0.

18. (C) 3

At each vertex of the triangle ABC draw a line parallel to its opposite side. These 3 lines intersect at 3 points labeled A', B', and C' as shown to the right. ABCB', BCAC', and CABA' are the only 3 parallelograms that are formed by adding a fourth point to the vertices of the triangle ABC.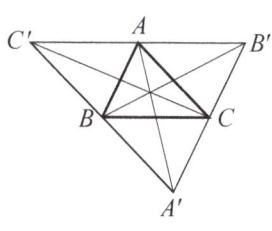

19. (D) 4

Each point is an endpoint of four different segments, so 4 must be assigned to any vertex that is connected to all three vertices with the 1, 2, and 3 that are already marked. There are four such vertices. They are labeled in the picture to the right. Altogether, there are 4 vertices with the label 4. The vertex with no label is connected only to the vertices with 4, so it can have a label 1, 2, or 3, but not 4.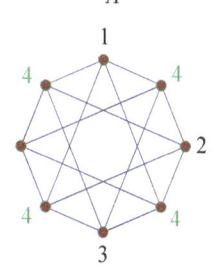

© Math Kangaroo in USA, NFP

20. (C) 3

3 pieces of candy would be left after the boy gives candy to the girls, so 80 − 3 = 77 pieces of candy would be given to the girls, the same number of pieces to each girl. The number of girls is at least 2 and at most 9 (because we already know there is at least one boy, the one who brought the candy), and it must be a divisor of 77. The only divisors of 77 are 1, 7, 11, and 77. Hence, the number of girls is 7 and the number of boys is 10 − 7 = 3.

21. (E) 20

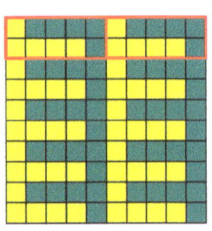

Each piece consists of 5 units, so any collection of such pieces consists of a multiple of 5 units and that multiple of 5 must be a perfect square. The smallest such multiples are 25 and 100. We can build a 10 × 10 square, as shown to the right. The first two rows consist of 4 original pieces, so the whole square requires 20 pieces.

To try to make a 5 × 5 square, we have 6 options to arrange one piece covering one corner unit. The options for the upper left corner are shown below.

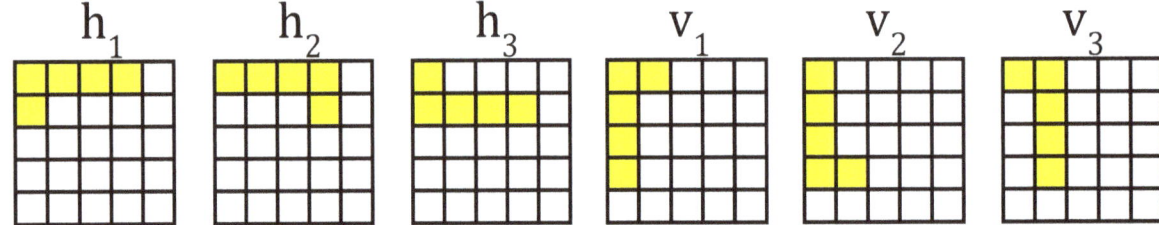

None of the arrangements cover any other corner unit or the center unit. This rule applies to all four corners, so any arrangement of 5 pieces covering a 5 × 5 square must have one piece at each corner and one piece at the center.

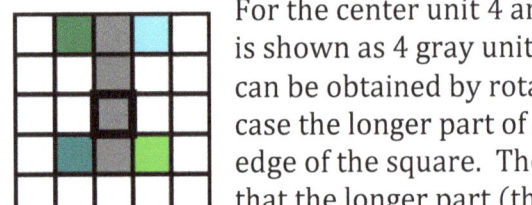

For the center unit 4 arrangements are shown to the left. Each arrangement is shown as 4 gray units plus one unit (in color). The other 12 arrangements can be obtained by rotating the square 90° or any multiple of 90°. In each case the longer part of a piece covers the center unit and is attached to one edge of the square. The cases are similar, so for an analysis we may assume that the longer part (the gray one) is situated as shown to the left. Below the same part is shown in red.

The upper left corner cannot be covered by any h-piece (where the longer part is horizontal) since it would have to cover part of the red center piece. Also, the v_2 position is excluded since it would not allow another piece to fit between the v_2 position and the red part. Hence, the upper left corner can be covered only by either v_1 or v_3 configuration.

In either case, the pieces at two left corners entirely cover the 2 leftmost columns. The same analysis works for the right corners, so the pieces at two right corners entirely cover the 2 rightmost columns.

In conclusion, there is no room for the 5th unit of the piece at the center unit, so a 5 × 5 square cannot be created.

Therefore, the smallest number of pieces Eve needed to make a square is 20 which forms a 10 × 10 square.

SOLUTIONS 2011

22. (E)

Figures (A) through (D) can be made as shown below.

(A) 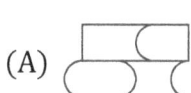 (B)  (C) 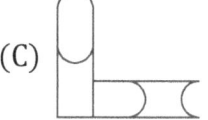 (D)

The top row of (E) is longer than the third of the four pieces shown 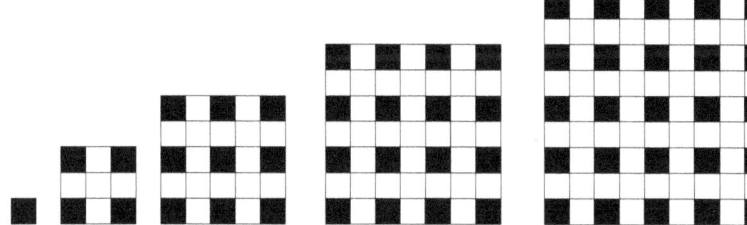, so at least the first and the third pieces are needed to cover the convex ends of the top row. After that, none of the other two pieces can be used to cover the convex end of the bottom row of the shape (E), so the shape (E) cannot be made out of the four pieces.

23. (C) 4
The smallest number among four selected numbers can be 2, 3, 5, or 6, since in each case we need at least 3 bigger numbers. 2, 3, and 5 are prime numbers, so if we pick one of them then the others 3 selected numbers must be multiples of 2, 3, or 5, respectively. It gives us three valid selections: {2, 6, 10, 30}, {3, 6, 15, 30}, and {5, 10, 15, 30}. There are only 3 numbers greater than 6, so if we pick 6 as the smallest number, the only option is {6, 10, 15, 30}. This is also a valid selection since for each pair of numbers there is a common factor greater than 1.

24. (D) 56
The 1 × 1 grid has 1 × 1 = 1 black square. The second grid (3 × 3) has 2 × 2 = 4 black squares. The third grid (5 × 5) has 3 × 3 = 9 black squares. According to this pattern, the grid with 25 black squares will be the fifth grid, because 5 × 5 = 25. The fifth grid will have the dimensions of 9 × 9, so it will consist of 81 squares total. Since it has 25 black squares, it will have 81 − 25 = 56 white squares.

You can also notice that the nth grid has the size of $(2n − 1) \times (2n − 1)$ and contains n^2 small black squares. Therefore, it contains $(2n − 1)^2 − n^2$ small white squares.
$(2n − 1)^2 − n^2 = (n − 1) \times (3n − 1)$.

© Math Kangaroo in USA, NFP   www.mathkangaroo.org

SOLUTIONS 2011

25. (B) 33 in

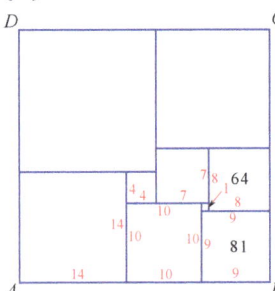

The squares with areas of 64 and 81 have side lengths of 8 and 9, respectively. The tiny square adjacent to the squares with side lengths 9 and 8 is a unit square since $9 - 8 = 1$. The other two squares adjacent to the unit square have the side lengths $8 - 1 = 7$ and $9 + 1 = 10$. These last two squares share a linear segment of a length $7 - 1 = 6$, so the other square adjacent to these two squares has the side length $10 - 6 = 4$. To the left of it is a square of a side length $4 + 10 = 14$. Hence, the length of side AB in inches is $9 + 10 + 14 = 33$. The other two squares have the side lengths 15 and 18 since $8 + 7 = 15$ and $4 + 14 = 18$.

26. (A) 5 Wednesdays.

If in a certain month there were 5 Saturdays and Sundays but only 4 Fridays and Mondays, then the 1st of the month was on a Saturday and the month had 30 days. The following month will have 31 days, because any month with 30 days is followed by a month with 31 days. The figure below shows that month and the following one. There will be 5 Mondays, 5 Tuesdays, and 5 Wednesdays in the following month. This corresponds only to answer (A) 5 Wednesdays.

Sun	Mon	Tue	Wed	Thu	Fri	Sat
						1
2	3	4	5	6	7	8
9	10	11	12	13	14	15
16	17	18	19	20	21	22
23	24	25	26	27	28	29
30	1	2	3	4	5	6
7	8	9	10	11	12	13
14	15	16	17	18	19	20
21	22	23	24	25	26	27
28	29	30	31			

27. (E) 6

The solid shown to the left consists of dice A, B, and C with t dots on the top face. The bottom face of A has $(7 - t)$ dots since for any die the number of dots on any two opposite faces add up to 7. The face of B that touches the $(7 - t)$ face has $[5 - (7 - t)] = t - 2$ dots since the sum of the number of dots on any two faces that touch is 5. By these rules, the opposite (bottom) face of B has $7 - (t - 2) = 9 - t$ dots and the face that touches it (the top face of C) has $5 - (9 - t) = t - 4$ dots.
The opposite face, the bottom of die C, has $7 - (t - 4) = 11 - t$ dots.
The six numbers t, $7 - t$, $t - 2$, $9 - t$, $t - 4$, and $11 - t$ all must be between 1 and 6.
$11 - 1 = 10$, $11 - 2 = 9$, $11 - 3 = 8$, and $11 - 4 = 7$ are all greater than 6, so t is either 5 or 6.
Die C already has 1 dot visible on its side face, so $t - 4$ can't be 1. Hence, t is not 5.
The only option left is $t = 6$, and the six numbers of the top and bottom faces of the three dice are 6, 1, 4, 3, 2, and 5, with 6 dots on the top face of the solid.

© Math Kangaroo in USA, NFP   110   www.mathkangaroo.org

SOLUTIONS 2011

28. (D) 6
Each pair of blue circles has exactly one point in common (the circles are externally tangent). The red circle and each blue circle have also exactly one point in common. These 6 common points are shown as intersection points of the 4 circles with 6 linear segments connecting the centers of the circles. The number of common points cannot be greater than the number of pairs of circles selected from the group of four circles.
For another example of 6 common points, we could use the big green circle and 3 blue circles internally tangent to it.
For any four objects A, B, C, and D, all the possible pairs are: AB, AC, AD, BC, BD, and CD. There are 6 pairs, so 6 is the greatest number of such points.

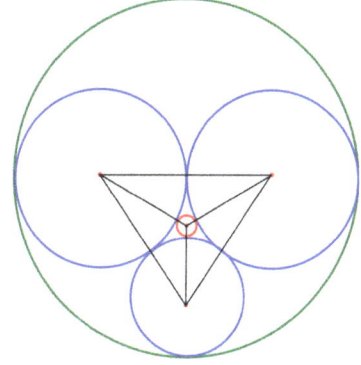

29. (C) 2
A chart of the boys accusing others of lying would look like this:
Tony → Alex → Thomas → Mark → Thomas
If Tony told the truth, then Alex was lying, so Thomas told the truth. Consequently, Mark was lying, and Thomas did tell the truth. In this case, there were 2 boys lying.
If, on the other hand, Tony was lying, then Alex told the truth, so Thomas was lying. Consequently, Mark told the truth confirming that Thomas was lying.
In this case, there were 2 boys lying. In either case, there were exactly 2 boys lying.

30. (A) There are no such numbers.
The last digit of this five-digit number has to be 5 for it to be divisible by 5. Also, the second and fourth digit would have to be even for those numbers to be divisible by 2 or 4, respectively. At this point we have only four such five-digit numbers:
12345 → 1234 is not divisible by 4,
32145 → 3214 is not divisible by 4,
14325 → 143 is not divisible by 3,
34125 → 341 is not divisible by 3.
Therefore, there are no such numbers.

Solutions for Year 2013

1. (E) 6

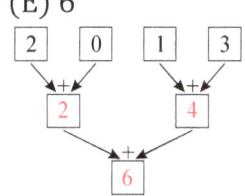
$2 + 0 = 2$, $1 + 3 = 4$, and $2 + 4 = 6$, so 6 is the answer.

© Math Kangaroo in USA, NFP   111   www.mathkangaroo.org

SOLUTIONS 2013

2. (C) 7

Look at the cube from the front left and consider the three vertical layers. In figure 2, the back layer has all the cubes. The middle layer is missing 2 cubes in the top row (shown in blue in the picture on the right). The front layer has 4 cubes, so it is missing 5 cubes (shown in green). So, 2 + 5 = 7 small cubes must be added.

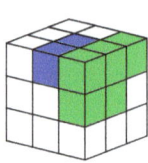

3. (C) 800 m

100 m is one-eighth of the distance from Mara to Bunica, so Mara covers 8 × 100 m = 800 m to get to Bunica.

4. (B) 4

Starting from A and moving to B, Nick has to make at least 4 right turns. The turns are marked by large dots. Nick could make his first right turn at the next intersection. This would not change the number of right turns.

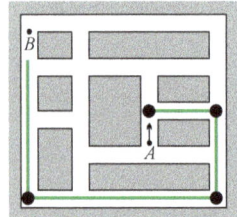

5. (E) 40

Three years from now the ages of Ann, Bob, and Chris will each increase by 3, so the sum of their ages will increase by 3 × 3 = 9. The sum of their ages will be 31 + 9 = 40.

6. (B) 4

176 as the product of a two-digit and a one-digit number can be written as 88 × 2, 44 × 4, or 22 × 8. Only 44 × 4 uses the same digit in each place.

7. (B) 11:50

Michael took his pills at 11:05, 11:20, 11:35, and his fourth pill at 11:50.

8. (E) 9

We get 8 regions by separating 8 corners of the squares and the 9th region as the common part of the two squares, so there are at most 9 regions.

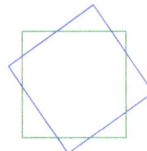

9. (C) 4

Dividing by 0 is not allowed, so both 20 and 30 are excluded from consideration. For a digit d, $20 + d$ is divisible by d only for $d = 1, 2, 4,$ and 5 (divisors of 20). There are 4 numbers between 20 and 30 such that the number is divisible by the digit in the ones position. The numbers are 21, 22, 24 and 25.

© Math Kangaroo in USA, NFP   112   www.mathkangaroo.org

SOLUTIONS 2013

10. (C) 4

The rectangle consists of 20 unit squares and each piece consists of 4 unit squares. There is more than one way to place 4 non-overlapping pieces in the 4 × 5 rectangle. If Ann tries to place 5 non-overlapping pieces in the rectangle, then she has to cover every unit (including the 4 corner units) of the rectangle since 5 × 4 = 20. The upper left corner unit can be covered only in the two ways shown to the left. For either choice there is only one way to cover the other three corner squares. After that, in either case, there is no room for the fifth piece. Therefore, 4 is the largest number of pieces Ann can put in the rectangle.

11. (C)

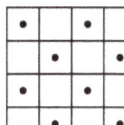

Look at the table. We move one unit at a time, either horizontally or vertically. The number of steps to move from a cell with a dot to another cell with a dot is always even. The same is true for cells without dots. The number of steps between a cell with a dot and a cell without a dot is always odd.

The same can be done for each of the five pieces shown above. Pick one open dot and make it red. Any open dot that you reach after any number of even steps is also colored red. The other dots are white. The picture to the left shows as an example the procedure for the piece in (A). The number of steps between two white dots is always even and number of steps between one red and one white dot is always odd. The largest number of either red or white dots is 4, 4, 5, 4, and 3, respectively, for (A), (B), (C), (D), and (E). Thus, the piece in (C) can cover 5 black dots of the table, which is the most of those listed. It actually can be done as shown to the right.

12. (C) 4

An easy way of seeing if the shape has the same perimeter as the square sheet of paper is to check whether the lines drawn inside of the square match up with the parts of the perimeter of the square not along the shape. The pictures below show the matching lines in red and purple, and the unmatched lines in black.

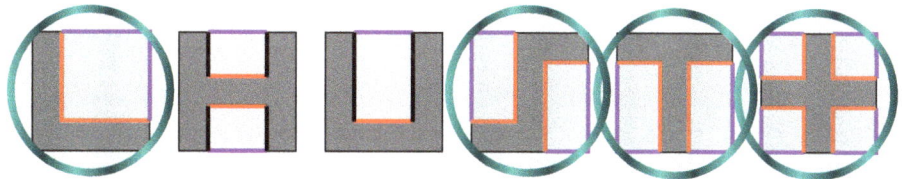

4 shapes have the same perimeters as the sheet of paper itself.

SOLUTIONS 2013

13. (D)

The first clock shows 1:30 p.m. and the second clock shows 3:30 p.m., so the ride takes her 2 hours. She is riding at a constant speed, so one-third of the ride takes one-third of the time. One-third of 2 hours is 40 minutes. 40 minutes after 1:30 p.m. is 2:10 p.m. This position of the minute hand at 10 minutes past the hour is shown in (D).

14. (B) 6
The difference between 3 × the fish caught and 1 × the fish caught is 12. This difference is 2 × the fish caught, so the number of fish caught is 12 ÷ 2 = 6.

15. (E)
When you look from the front of the building you will see the highest tower in each column. These heights are represented by the largest number in each column of the table. The numbers are 4, 3, 3, and 2, respectively. (E) shows these heights.

16. (B) 8 or 9
Besides the winner and the last place finisher, the three candidates received 36 – 12 – 4 = 20 votes. The votes for these candidates are different and between 5 and 11. If the fourth place finisher got 6 or more votes, then the sum of votes for the three candidates would be at least 6 + 7 + 8, which exceeds 20. Hence, the fourth place finisher got 5 votes and the other two got a total of 15 votes, each with between 6 and 11 votes. The only two options are 6 + 9 or 7 + 8, so the candidate in the second place got either 8 or 9 votes.

17. (D) 30
Cutting out one small cube at a corner of the big cube adds 3 new faces to the solid, so cutting out one small cube at all 8 corners adds 8 × 3 = 24 new faces to the original 6 faces of the big cube. The final solid has 6 + 24 = 30 faces.

18. (A) 40
All numbers from 10 to 99 are two-digit numbers. The bigger of the numbers in each pair has to be at least 10 more than 50, so the first pair we will count is 60 – 10 = 50. The largest two-digit number is 99, so the last pair we will count is 99 – 49 = 50. There are 40 numbers from 60 to 99 inclusive (and from 10 and 49), so there are 40 such pairs.

19. (C) 5
Six goals were scored in the first half, with the guest team leading at that point, so the result could have been 0 : 6, 1 : 5, or 2 : 4. However, the home team scored three goals in the second half and won the game, so the difference between the numbers of goals scored by the two teams in the first half could not have been more than 3. So, the only result at half-time could have been 2 : 4 (2 goals for the home team and 4 goals for the guests). Therefore, the final result of the championship match was 5 : 4 for the home team.

SOLUTIONS 2013

20. (D) 7

By the ±1 rule for adjacent cells, the highest number in a red cell can be 4, the highest number in a blue cell can be 5, the highest number in a green cell can be 6, the highest number in a purple cell can be 7, and the highest number in an orange cell can be 8. Therefore, 9 can be only in the lower right corner. Since 9 must be in the table, it occurs in the lower right corner cell.

Move back from 9 to 3 and follow the ±1 rule for adjacent cells. The lowest number in the orange cells can be 8, the lowest number in the purple cells can be 7, the lowest number in the green cells can be 6, the lowest number in the blue cells can be 5, and the lowest number in the red cells can be 4. For each cell color the highest and the lowest numbers are the same, so there is only one solution for this puzzle and it is shown to the right. 7 different numbers (all the numbers from 3 to 9 inclusive) appear in the table.

21. (A) Aron's stone is green.

Aron is lying, so Aron and Bern have stones of different colors. Bern is lying, so Bern and Carl have stones of different colors. There are only two colors, so Aron and Carl have stones of the same color and Bern's stone is the other color. Exactly two boys have stones of the same color but it is not red since Carl is lying. In summary, Aron's stone is green, Bern's stone is red, and Carl's stone is green. Also, Aron's stone and Carl's stone are the same color. Therefore, only "Aron's stone is green" is a true statement.

22. (D) 14

After the first round there were 66 − 21 = 45 cats remaining. 27 of the 45 cats had stripes and 32 had one black ear. If none of the 45 cats were both striped and with one black ear, there would have been 59 cats remaining, which is 14 more than the actual number of 45, so at least 14 were striped cats with one black ear. Notice that 14 = (27 + 32) − 45. Since all these cats made it to the final, there were at least 14 finalists.

SOLUTIONS 2013

23. (B) 3
Let H be a button with a happy face and S one with a sad face. We start with SHSH and press just one of the four buttons. The outcomes are displayed below.

1st button	2nd button	3rd button	4th button
HSSH	HSHH	SSHS	SHHS

None of them is HHHH, so in each case we press one button again (it could be the same button but pressing the same button again would return us to the original SHSH).
The table below displays all outcomes of pressing two buttons. The button with the number at the left margin is pressed first and the button with the number at the top margin is pressed second.

	1	2	3	4
1	SHSH	SHHH	HHHS	HSHS
2	SHHH	SHSH	HHSS	HSSS
3	HHHS	HHSS	SHSH	SSSH
4	HSHS	HSSS	SSSH	SHSH

None of the outcomes are HHHH, so in each case we need to press one button again. The outcomes in green can be converted into HHHH after the 3rd press of the button. Thus, we have to press the buttons at least 3 times to get all happy faces.
Notice this works only if we never press the 1st button and when all three buttons that we press are different.

24. (A) 18
To figure out how many boys give their left hand to a girl when we know the number of boys who give their right hand to a girl, let's begin with any girl in a circle. Move to her right until we get to a boy and move to her left until we get to a boy. There could be multiple girls in this chain or there could be just one girl. A boy at the right end of the chain holds a girl by his left hand and a boy at the left end of the chain holds a girl by his right hand. Thus, the number of boys who give their left hand to a girl equals the number of boys who give their right hand to a girl. Therefore, 18 boys give their left hands to a girl.

SOLUTIONS 2013

25. (D) 7
4 black and 4 white unit cubes are used to form our $2 \times 2 \times 2$ cube. The seven basic arrangements are displayed below.

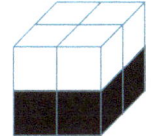 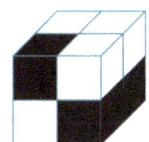 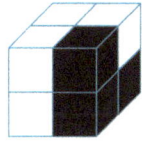 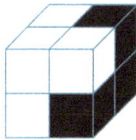 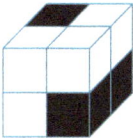 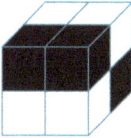 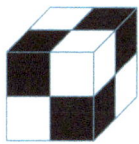

For each of the 7 solids the invisible back lower left corner is a black unit cube. The bottom of the first solid is black. Each of the next four solids has one white unit cube at the bottom. Each of the last two solids has two white unit cubes at the bottom. The plans for the seven arrangements are displayed below.

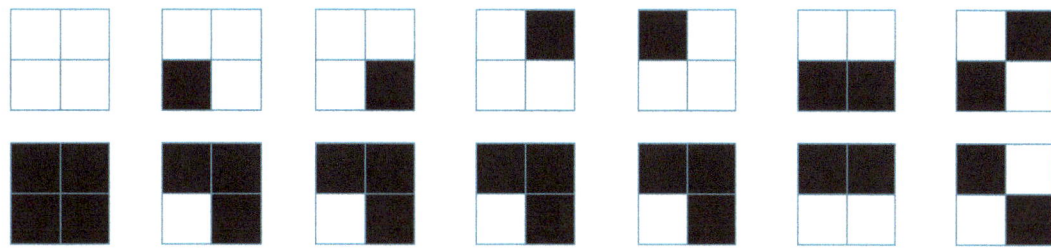

The top is shown above the corresponding bottom. All other arrangements for the $2 \times 2 \times 2$ cube are obtained by rotating the above solids.

26. (D) 60
Let abc and cba be 3-digit numbers such that their difference is 297. Neither a nor c is 0. The difference $(100a + 10b + c) - (100c + 10b + a) = 99a - 99c = 99(a - c)$ is 297, so $a - c = 3$ since $297 \div 99 = 3$. a and c are non-zero digits, so a can be any digit from 4 to 9, c is completely determined by a, and b can be any digit from 0 to 9. There are $6 \times 10 = 60$ such numbers (6 options for a and 10 options for b). The smallest among them is 401 and the largest is 996. Indeed, $401 - 297 = 104$ and $996 - 297 = 699$.

27. (B) 12
Matthew made a perfect circle from 8 identical track parts, so 4 tracks make a semi-circle. Marten can insert two sets of tracks that he started with (4 pieces altogether) between the two separated semi-circles from Matthew's track. Marten can use as few as 12 track parts to make a closed track (shown).

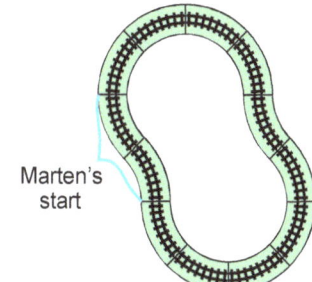

SOLUTIONS 2013

28. (B) 1006

In the reverse order look at the islanders making the same statement, "After my departure the number of knights on the island will equal the number of liars." Call them #2013, #2012, #2011, #2010, ⋯, #4, #3, #2 and #1. Compare the number of knights to the number of liars. After #2013's departure, there was nobody left. $0 = 0$ is true, so #2013 was a knight. After #2012's departure there was 1 knight (#2013) and no liars. $1 = 0$ is false, so #2012 was a liar. After #2011's departure, there was 1 knight (#2013) and 1 liar (#2012). $1 = 1$ is true, so #2011 was a knight. #2013 and #2011 were knights and #2012 was a liar. $2 = 1$ is false, so #2010 was a liar. #2013 and #2011 were knights, and #2012 and #2010 were liars. $2 = 2$ is true, so #2009 was a knight. #2013, #2011, and #2009 were knights, and #2012 and #2010 were liars. $3 = 2$ is false, so #2008 was a liar. #2013, #2011, and #2009 were knights, and #2012, #2010, and #2008 were liars. $3 = 3$ is true, so #2007 was a knight. #2013, #2011, #2009, and #2007 were knights, and #2012, #2010, and #2008 were liars. $4 = 3$ is false, so #2006 was a liar.

Continue to see that the islanders with odd numbers were knights and the islanders with even numbers were liars. There are $2012 \div 2 = 1006$ even numbers from #2012 to #2, so initially there were 1006 liars. Also, since we figured out that #1 was a knight, after his departure there were 2012 people and half of them were liars. Thus again, the number of liars was $2012 \div 2 = 1006$.

29. (D) 19

Let's apply the "change-sum" procedure a few times to {20, 1, 3} and compute the algebraic differences between the terms of the triples.

Terms	{20, 1, 3}	{4, 23, 21}	{44, 25, 27}	{52, 71, 69}	{140, 121, 123}	{244, 263, 261}
1st – 3rd	+19	-19	+19	-19	+19	-19
2nd – 3rd	-2	+2	-2	+2	-2	+2
3rd – 1st	-17	+17	-17	+17	-17	+17

Each row has the same number with an alternating sign. When we talk about the difference between two numbers, we mean the algebraic difference without the sign (we subtract the smaller of two numbers from the bigger one), so the maximum difference between two numbers of the list after 2013 consecutive "change-sums" is 19 if the pattern holds.

Well, the alternating sign is not a coincidence and it doesn't depend on an initial list. Pick any list of three numbers, let's say {7, 13, 5}. (Also try your own list.) Apply the "change-sum" procedure to get {13 + 5, 5 + 7, 7 + 13}, which is {18, 12, 20}. Add the three numbers from the initial list: $7 + 13 + 5 = 25$. Clearly, $18 = 25 - 7$, $12 = 25 - 13$, and $20 = 25 - 5$ since $(7 + 13 + 5) - 7 = 13 + 5$, $(7 + 13 + 5) - 13 = 5 + 7$ and $(7 + 13 + 5) - 5 = 7 + 13$.

The algebraic difference (1st – 2nd) for the second list is $18 - 12 = 6$.

$18 = 25 - 7$ and $12 = 25 - 13$, so $18 - 12$ can be computed as $(25 - 7) - (25 - 13) = 25 - 7 - 25 + 13 = -7 + 13 = -(7 - 13)$. Thus, $18 - 12 = -(7 - 13)$ where $7 - 13$ is the (1st – 2nd) algebraic difference for the initial list, so the two corresponding algebraic differences for the initial and second lists are the same except for the opposite signs. The same calculation works for the algebraic differences (2nd – 3rd) and (3rd – 1st). Hence, the "change-sum" procedure never changes the maximum difference after any number of applications for any initial list. For our particular list {20, 1, 3} the maximum difference is 19 and it doesn't change after 2013 applications.

SOLUTIONS 2013

30. **(B) 68**
The largest total of all the numbers on the outside surface of the block that Alice can get will happen if the identical numbers that are glued together will have the smallest values possible. Typically, we would pick only faces with 1 and faces with 2 to be glued to the faces with the same number. However, in the net that Alice is using faces with 1 and 2 are on the opposite sides, whereas the glued parts of the block have to be adjacent. Therefore, Alice would glue 1 and 3 on the inside. Since she can flip the cubes, this is possible. This arrangement leaves 2, 4, 5, and 6 on the outside of the block, so the largest total on the outside surface of the block is $4 \times (2 + 4 + 5 + 6) = 68$.

Solutions for Year 2015

1. (B)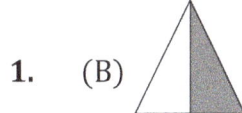

   Each of the figures is divided into equal parts. The fraction under each figures indicates how much of its area is shaded.

   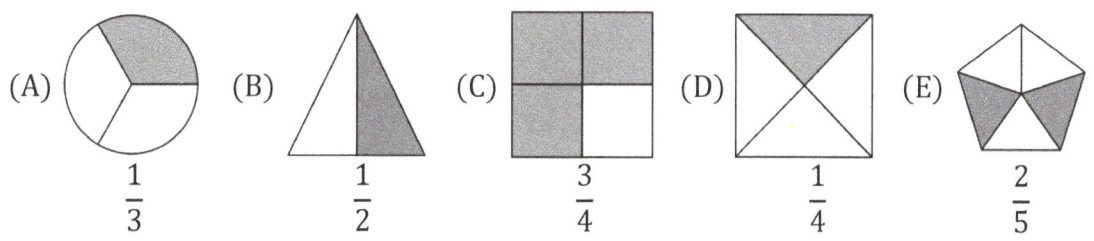

   Only figure (B) has one half of its area shaded.

2. (C)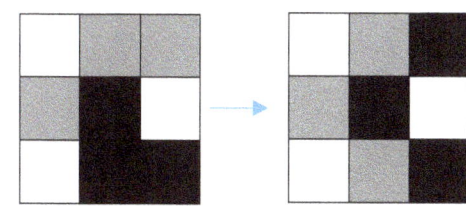

   Rotate the umbrella 180° to see that Я doesn't fit; it is backwards. The other four umbrellas: ANG, OOK, NGA, and ARO, are parts of the big umbrella.

3. (A) 2
   In the top row there are two gray squares with a common side and in the bottom row there are two black squares with a common side, so it is not enough to repaint just 1 square. After repainting 2 squares, the one in the top right to black and the bottom middle to gray, as shown to the right, no two squares with a common side are the same color.

SOLUTIONS 2015

4. (A) 75

5 ducks lay an egg every day, so they lay 5 × 10 = 50 eggs in a period of 10 days. The other 5 ducks lay an egg every other day, so they lay 5 × 5 = 25 eggs in a period of 10 days. Altogether, the 10 ducks lay 50 + 25 = 75 eggs in a period of 10 days.

5. (B) 18 cm

The area of each small square is 4 cm², so all sides of the small squares have a length of 2 cm. The thick black line covers 9 sides, so its length is 9 × 2 cm = 18 cm.

6. (E) $\frac{23}{12}$

Here is a simple arithmetic for all five fractions.
(A) $\frac{19}{8} > \frac{16}{8} = 2$
(B) $\frac{20}{9} > \frac{18}{9} = 2$
(C) $\frac{21}{10} > \frac{20}{10} = 2$
(D) $\frac{22}{11} = 2$
(E) $\frac{23}{12} < \frac{24}{12} = 2$
Only $\frac{23}{12}$ is smaller than 2.

7. (D) 5 kg

If we replace the watermelon by one pumpkin, then the combined weight will increase by 2 kg from 8 kg to 10 kg. Thus, 2 pumpkins weigh 10 kg and one pumpkin weighs 5 kg.

8. (A) 10

The flowers are only on one kind of plants, and there is only 1 flower on each of these plants. We need 6 such plants to have a total of 6 flowers. These 6 plants have 6 × 2 = 12 leaves. We need 32 − 12 = 20 more leaves and they are on plants with 5 leaves only. There are 20 ÷ 5 = 4 such plants, so together there are 6 + 4 = 10 plants.

9. (A) 4 cm

For any two strips, the sum of lengths of two original unglued strips is the same as the sum of lengths of the glued strip and the overlap. Also, this means that the overlap is the difference between the sum of the lengths of two original strips and the length of the glued strip. For two strips the overlap is 10 cm long and the glued strip has the length of 50 cm, so the sum of the lengths of the two original unglued strips is 50 cm + 10 cm = 60 cm. If we want the glued strip to be 56 cm long, then the overlap must be 60 cm − 56 cm = 4 cm long.

10. (D) 12

The perimeter of the shape consists of 6 vertical units, 1 top unit, 3 bottom units, two short horizontal red segments above the bottom row and two short horizontal blue segments above the middle row. The total length of the red segments is 3 − 2 = 1 and the total length of the blue segments is 2 − 1 = 1. Thus, the perimeter is 6 + 1 + 3 + 1 + 1 = 12.

SOLUTIONS 2015

11. (E) 20
Numerically, the months from January to September are labeled 01 to 09, and the months from October to December are labeled 10 to 12. The largest sum of digits for the months is $0 + 9 = 9$. The days can be labeled from 01 to 31, and the largest sum for the days is $2 + 9 = 11$. September 29 is a calendar date, so $0 + 9 + 2 + 9 = 20$ is the largest sum calculated by Mary.

12. (C) 2 cm

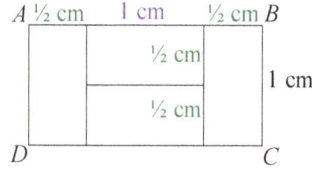

For the 4 equal rectangles, the two short sides are the same length as one long side, so each long side is 1 cm long and each short side is ½ cm long. The length of AB is ½ cm + 1 cm + ½ cm = 2 cm.

13. (C)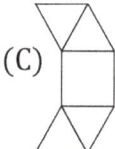

In each case, the square is the base of the pyramid, and the four triangles should form the lateral faces. In the net shown in (C), when folding the two triangles adjacent to the square so that their vertices meet over the center of the base, we can see that the other two triangles completely overlap along the left edge of the base. So, the face in the pyramid along the right edge of the base will be missing. The other four diagrams can be easily folded into the pyramid.

14. (C) 4030
If somebody has already had their birthday in a given year, the sum of their age and birth year will be the current year. Lucy and her mother were born in February, so on March 19, 2015, the sum of Lucy's age and birth year is 2015. The same is true for her mother, so together the numbers add up to $2015 + 2015 = 4030$.

15. (C) 5
$2 \times (7 + \text{multiple of } 9)$ is $14 + \text{multiple of } 9$. $14 = 5 + 9$, and $(5 + 9) + \text{multiple of } 9$ is $5 + \text{another multiple of } 9$, so the remainder is 5.

16. (B) 26
The three possible dimensions of the rectangle are 1×12, 2×6, and 3×4. The corresponding perimeters are $2 \times (1 + 12) = 26$, $2 \times (2 + 6) = 16$, and $2 \times (3 + 4) = 14$. Only 26 is among the options listed.

SOLUTIONS 2015

17. (C) only red

Let's label the triangles with A, B, C, and D. The common side of triangles A and B is neither blue nor red, so it must be green.

Triangle B now has a green side and a red side, so its third side must be blue.

The common side of triangles C and D is neither blue nor red, so it must be green. The third side of triangle C (the side with *x*) must be red.

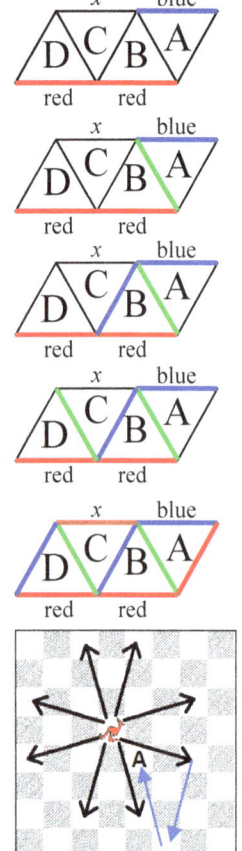

18. (E) 13

If Simon has perfectly bad luck, then he can possibly take out all 5 yellow apples and all 7 green pears without yet having taken out at least one apple and one pear of the same color.

After taking out these 12 pieces of fruit, he can pick any of the 5 pieces left (3 green apples, 2 yellow pears) as his 13th piece of fruit. If the 13th piece is a green apple, then he has a one green apple and one green pear. If the 13th piece is a yellow pear, then he has a one yellow apple and one yellow pear.

19. (B) 3

None of the places the kangaroo can immediately move to are one move away from A. After the first move there are some places that are two moves away from A. One of these paths is shown in the figure, so the minimum number of moves the kangaroo needs to make is 3.

20. (E) 6

$X + X + YY$ is at most $8 + 8 + 99 = 115$, so Z must be 1 and the sum is 111. YY cannot be 88 (or less) since $111 - 88 = 23$, which is always more than $X + X$ no matter what digit X stands for. Hence YY must be 99. The difference $111 - 99$ is 12 and the only solution for $X + X = 12$ is $X = 6$.

X represents 6, Y represents 9, and Z represents 1. These are all different digits.

21. (C) 16

80 as a product of two natural numbers can be expressed in 5 different ways: 1×80, 2×40, 4×20, 5×16, and 8×10. We can't use 1×80 or 2×40 since any sum of four natural numbers containing 80 or 40 is greater than 39. 39 is an odd number, so the sum of the four numbers must contain at least one odd term to be equal to 39. Among the remaining products only 5×16 has an odd factor, so the numbers 5 and 16 must appear in the sum. $5 + 16 = 21$, so the remaining two factors need to add up to $39 - 21 = 18$. Of the pairs we found, only $10 + 8 = 18$, so the four numbers are 5, 16, 8, and 10. The largest of these four numbers is 16.

SOLUTIONS 2015

22. (C) 34 dollars

After purchasing the first two books Jane was left with a certain amount of money. This amount of money was spent on the third book and was equal to [half of it + $3]. Thus, this book cost $6, because half of the amount of money she had at this point was $3, and this was the amount of money left after she purchased the second book.

The amount of money left after purchasing the first book was spent on the last two books and was equal to [half of it + $2] (spent on the second book) + $6 (spent on the third book). This money equals [half of it + $8]. Hence, she had $16 left after purchasing the first book. Jane spent $8 + $2 = $10 on the second book.

The $16 + $1 + half of Jane's original money is exactly Jane's original money, so before any purchase Jane had 2 × ($16 + $1) = $34 and spent the whole $34 on the three books. She spent $34 ÷ 2 + $1 = $17 + $1 = $18 on the first book.

She paid $18 for the first book, $10 for the second book, and $6 for the third book.

23. (D) 29

We can start off by drawing the 9 houses

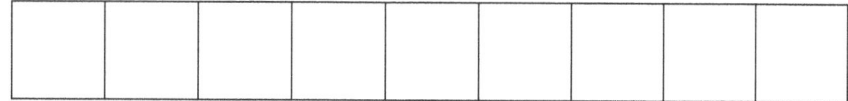

Since we know that each house has at least 1 person, and any 2 adjacent houses have at most 6 total people, this means that a single house can at most have 5 people, and the adjacent house will have 1 person. Since there is an odd number of houses, if we want to maximize the number of people, then we should have the 5 person houses on the outside.

| 5 | 1 | 5 | 1 | 5 | 1 | 5 | 1 | 5 |

This will give us the total 29.

24. (C) 67

2 × 100 or 3 × 100 are multiples of 100 and are therefore already divisible by 4. Adding 1 or 2 and then dividing by 4 will not result in a natural number, so in the third operation we can't divide by 4.

3 × 100 is already divisible by 3. Adding 1 or 2 and then dividing by 3 will not result in a natural number, so in the first operation we can't multiply by 3.

When first multiplying by 2, adding 1 or 2, then dividing by 3 the results are
((100 × 2) + 1) ÷ 3 = 201 ÷ 3 = 67 or ((100 × 2) + 2) ÷ 3 = = 202 ÷ 3 = 67⅓. The correct operations to get a natural number as the result is to first multiply by 2, then add 1, and finally divide by 3 to obtain 67 as the answer.

SOLUTIONS 2015

25. (B) 61

$\overline{abcd}$ is a 4-digit number with digits in increasing order. $\overline{bd} - \overline{ac} = (10b + d) - (10a + c) = 10(b - a) + (d - c)$. The largest possible value for $b - a$ is $7 - 1 = 6$ since $a > 0$ as the first digit of a 4-digit number and $b < c$ where $c \leq 8$ since d has to be the greatest digit.
$b - a = 6$ only if $a = 1, b = 7, c = 8,$ and $d = 9$. In this case $\overline{bd} - \overline{ac} = 79 - 18 = 61$.
If $(b - a) \leq 5$, then $10(b - a) + (d - c) \leq 50 + (d - c)$ and $50 + (d - c) < 60$ for any digits d and c. Thus, $\overline{bd} - \overline{ac}$ is less than 60 if $b - a$ is not 6, so the largest value for $\overline{bd} - \overline{ac}$ is 61.

26. (C) 22

Vertices D and C share two faces along the edge DC, so $16 - 14 =$
= the number on EHDA (the left face) – the number on FGCB (the right face).
Similarly, vertices E and F share two faces along the edge EF, so
(24 – the number at F) is the difference between the numbers on the left and right faces, which is 2. (24 – the number at F) = 2, so the number at F is 22.

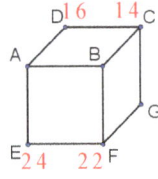

27. (B) 8

Let n be the number of compartments in one car. There are $6n$ compartments in the first six cars. The compartments in the 7th car are $6n + 1, 6n + 2, 6n + 3, ..., 7n$. 50 is between $6n + 1$ and $7n$ if and only if $n = 8$, so there are 8 compartments in each car.

28. (E) 10

All 10 combinations are shown below.

© Math Kangaroo in USA, NFP

SOLUTIONS 2015

29. (E) 9

Let A, B, C, D be the four points from left to right. Then AB, BC, and CD are each less than 11. Indeed, if one of them is 11 or more, the other two are at least 2 and 3, so AD = AB + BC + CD ≥ 2 + 3 + 11 = 16 > 14 (the longest segment).
Thus, AB, BC, CD are equal to, in some order, 2, 3, and k. Hence, 14 = AD = AB + BC + CD, which is $2 + 3 + k = 5 + k$. Thus, $5 + k = 14$ and $k = 9$.

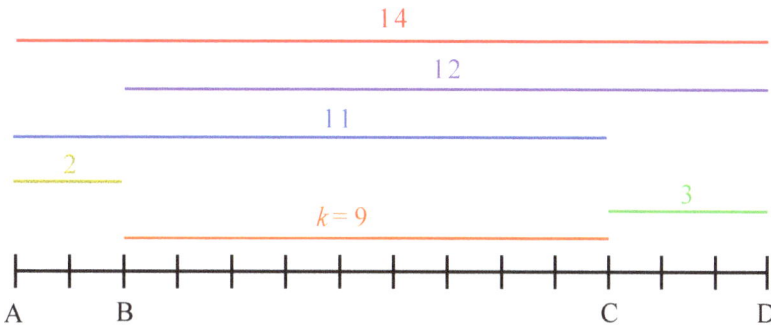

30. (D) 24

Any vertex of a big cube is also a vertex of a corresponding small cube, so the three faces of the big cube that share a vertex cannot all be red. If the three faces were blue, then the three faces at the opposite vertex would be red. Therefore, we need two colors at each of the 8 vertices of the big cube. Switching colors is not changing the number of small cubes containing two colors, so we can assume (rotating the big cube if necessary) that the front face and the left face are red. Then the top and bottom faces must be blue. Among the other two faces one is red and one is blue. For example, the right face is blue and the face behind is red. On the right face, 2 horizontal edges are blue, because the faces sharing the edges are blue. On the left face, the 2 vertical edges are red. Other edges share faces of different colors. Every cube has 12 edges, so our big cube has 12 − 4 = 8 edges with two colors. Not counting the 8 small cubes at the vertices of the big cube, each one-color edge yields only one-color cubes and each two-color edge yields 2 two-color cubes, so the number of all cubes with two colors is 24: 8 at the vertices + 8 × 2 along the two-color edges + 4 × 0 along the one-color edges.

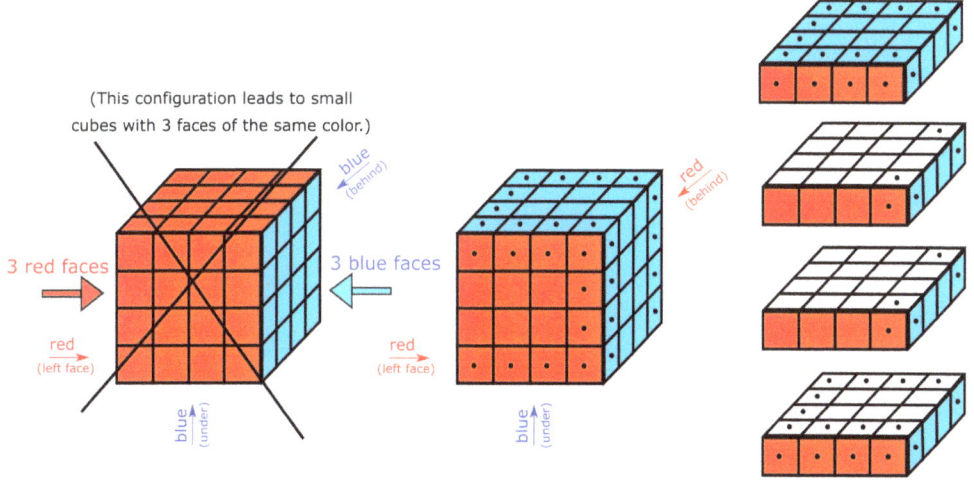

© Math Kangaroo in USA, NFP   125   www.mathkangaroo.org

Solutions for Year 2017

1. (B)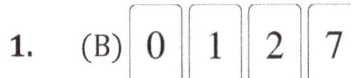

    (B) has three cards out of their original positions, so it could not have been made with only one swap of two cards. For the other cases the swaps are shown below.

    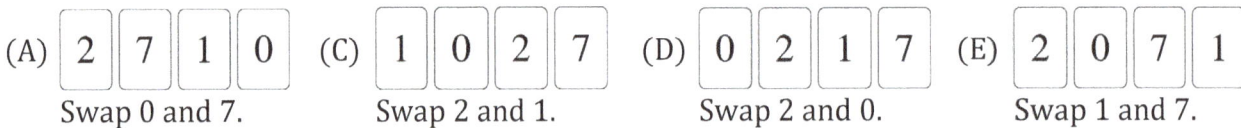

2. (C) 4 cats.
    3 flies and 2 spiders have $3 \times 6 + 2 \times 8 = 34$ legs. 9 chickens have $9 \times 2 = 18$ legs. The difference is 16 legs, and 16 legs belong to 4 cats since $4 \times 4 = 16$.

3. (E)

    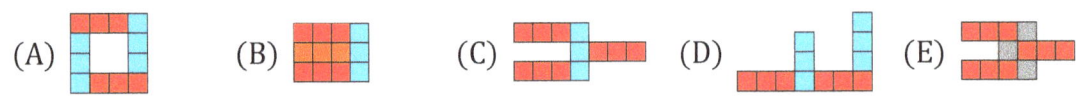

    Possible solutions are shown above. In (E), the leftmost and the rightmost cells need to be covered by horizontal pieces (shown in red), but the gray area cannot be covered by ▭▭▭.

4. (D) 2468642
    $1111 \times 2222 = 1111 \times 1111 \times 2 = 1234321 \times 2 = 2468642$.

5. (B) 2
    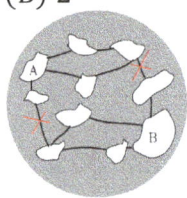
    There are at least two disjoint paths connecting A and B, so at least two bridges must be closed for traffic. Two is enough as shown by the example in the diagram.

6. (A)

    The walking order is: Jane – Kate – Lynn. Jane is heavier than Kate by 500 kg and Lynn is heavier than Kate by 1000 kg, so Kate is the lightest and Lynn is the heaviest among the three, as depicted in (A).

SOLUTIONS 2017

7. (E) 15

If four different numbers are split into two pairs with equal sums, then one pair consists of the smallest and the largest numbers. The other pair consists of the two middle numbers. From the list 5, 6, 9, 11 and 14 drop one number (which is on the face opposite the sixth face) and then for the list of four numbers check the sum of its middle numbers against the sum of the other two numbers.
When you drop 14, then $(5 + 11) \neq (6 + 9)$. When you drop 11, then $(5 + 14) \neq (6 + 9)$.
When you drop 9, then $(5 + 14) \neq (6 + 11)$. When you drop 6, then $(5 + 14) \neq (9 + 11)$.
Drop 5 and see that $(6 + 14) = 20 = (9 + 11)$ and the number on the sixth face is $20 - 5 = 15$.

8. (C) 3

$\frac{1}{3}$ of the squares will be blue and $\frac{1}{2}$ of the squares will be yellow, so
$1 - \left(\frac{1}{3} + \frac{1}{2}\right) = 1 - \left(\frac{2}{6} + \frac{3}{6}\right) = 1 - \frac{5}{6} = \frac{1}{6}$ of the squares will be red.
The rectangle contains 3×6 squares, so Martin will color
$\frac{1}{6} \times 3 \times 6 = \frac{1}{6} \times 6 \times 3 = 1 \times 3 = 3$ squares red.

Or: There are $3 \times 6 = 18$ squares. $18 \div 3 = 6$ squares will be colored blue and $18 \div 2 = 9$ squares will be colored yellow. This leaves 18 - (6 + 9) = 18 - 15 = 3 squares to be colored red.

9. (B) 6

Split 30 using the ratio of 2 to 3, which is $\frac{2}{2+3} \times 30$ to $\frac{3}{2+3} \times 30$, or 12 to 18. Thus, Nick solved 18 problems and Peter solved 12 problems. $18 - 12 = 6$, so Nick solved 6 more problems.

10. (D)

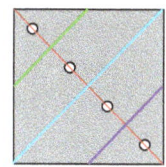

We need either two or three axes of symmetry perpendicular to the line with the four holes (shown in red). The blue line is the 1st axis of symmetry (the first folding line), so after the first folding we see only two holes. Then the green line becomes the 2nd folding line, so after the second folding we see only one hole. Bob also could have folded the paper three times, first folding in the corners along the green and purple lines, and then folding along the blue line. Figure (D) shows these lines, no matter which of the two ways Bob did the folding.

11. (D) 100 cm

The difference between the width of the sofa and the width of the loveseat is the width of one white piece, so the width of one white piece is 220 cm – 160 cm = 60 cm. Looking at the loveseat, the width of two armrests is 160 cm – 2 × 60 cm = 40 cm. The width of the chair is 40 cm (two armrests) + 60 cm (one white piece) = 100 cm.

SOLUTIONS 2017

12. (C) 284
Only DAD contains two identical letters, so 414 is the key that fits it. HAB is the only other padlock with A in the middle, so its key is 812. So far, 1 = A, 2 = B, 4 = D, and 8 = H. ABD corresponds to 124 and AHD corresponds to 184. Thus, the key with the question mark fits the padlock with BHD, and its number is 284.

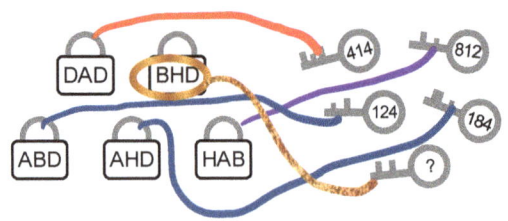

13. (C) 9781920
After deleting 24 digits from 1234567891011121314151617181920, the largest numbers among 7-digit (31 − 24 = 7) numbers have to start with 9. We can't start with the last 9 of the 31-digit number as that would mean that more than 24 digits in front of it were deleted, so we have to start with the first 9. We want the next digit to be as large as possible. It can't be the last 9 and it can't be the last 8 since both 9920 and 981920 have less than seven digits. The best option is 7 as the next digit. After this 7 there are still six digits left, 181920. For the largest number we drop 1 in front of 8, so 9781920 is our answer.

14. (C) $3 \times 4 \times 5$

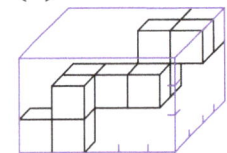

By counting along the height (3 levels), along the depth (4 levels), and along the width (5 levels) we can see that the dimensions of the smallest possible box are $3 \times 4 \times 5$.

15. (D) a is less than d.
The equations are: $a + b = 2, c + d = 3, a + c = 1,$ and $b + d = 4$.
$a + b = 2$ and $a + c = 1$, so $b = c + 1$, which excludes statements (B) and (E).
$a + b = 2$ and $b + d = 4$, so $d = a + 2$. Therefore, a is less than d, which is statement (D). This excludes statements (A) and (C).

16. (E) 16 km
Wednesday is in the middle of the 5-day period, so on Tuesday Peter walked **2 km less** than on Wednesday and on Thursday he walked **2 km more** than on Wednesday. On Tuesday and Thursday together Peter walked twice the distance he walked on Wednesday. The same is true for Monday and Friday. Hence, during the 5-day period Peter walked a distance 5 times as long as the distance he walked on Wednesday. Thus, on Wednesday he walked the distance of 70 km ÷ 5 = 14 km. On Thursday Peter walked 14 km + 2 km, which is 16 km.

17. (E) Step-by-step reflections are shown below.

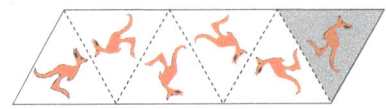

18. (D)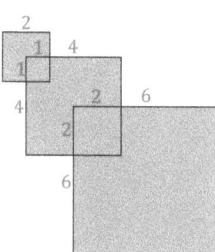

The options (A), (B), (C), (D), and (E) are translated as follows: (A) ×2, +1, −1; (B) +1, −1, ×2; (C) ×2, −1, +1; (D) +1, ×2, −1; and (E) −1, +1, ×2. Let b be Boris' initial amount of money. Then each option results in: (A) [(b × 2) + 1] − 1 = 2b; (B) [(b + 1) − 1] × 2 = 2b; (C) [(b × 2) − 1) + 1 = 2b; (D) [(b + 1) × 2] − 1 = **2b + 1**; and (E) [(b − 1) + 1] × 2 = 2b. Each option doubles Boris' money, but option (D) doubles it and adds 1 more dollar.

19. (B) 51 cm²
The original squares have the areas equal to 2^2, 4^2, and 6^2 (in cm²).
The middle square overlaps with the small square and the big square.
The parts which overlap are disjoint squares with the areas 1^2 and 2^2, so the shaded area is:
$(2^2 + 4^2 + 6^2) − (1^2 + 2^2) = (4 + 16 + 36) − (1 + 4) = 56 − 5 = 51$ cm²

20. (C) 4
Mike's score is less than the lowest score among the other three players who scored 20 goals in total. The lowest score among the three players can't be 6 since 6 + 7 + 8 is more than 20 but the lowest score among the three players can be 5 since 5 + 6 + 9 = 20 or 5 + 7 + 8 = 20. Hence, the largest number of goals Mike could have sored is 4.

SOLUTIONS 2017

21. (A)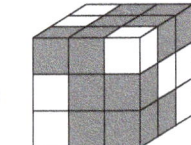

To construct (A) place all 9 bars perpendicularly to the right face. Each bar can have the white end either to the left or to the right. The direction is indicated by what we see on the right face. It is relatively easy to prove that (B), (C), and (D) cannot be constructed by using 9 bars. The width direction is the direction perpendicular to the right face, the depth direction is the direction perpendicular to the front face, and the height direction is the direction perpendicular to the top face.

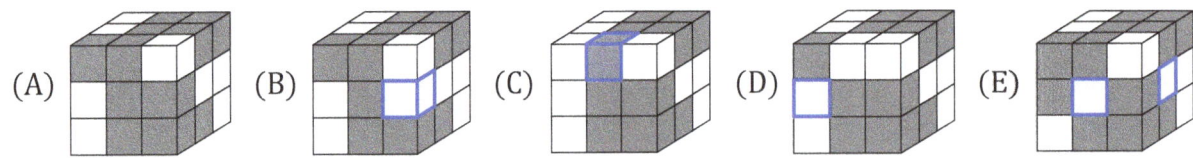

For (B), focus on the middle white cube of the common edge of the front face and the right face. If you move from it in any of the three directions, you see two or three white cubes, so the middle cube can't belong to any bar since each bar has only one white cube at one of its ends. For (C), focus on the gray middle cube of the common edge of the front face and the top face. In two of the three directions you can see only gray cubes and in the width direction you see two white cubes, so the middle cube can't be a part of any bar. For (D), focus on the gray left corner cube of the common edge of the front face and the top face. In either of the three directions you can see two white cubes, so the left corner cube can't be a part of any bar. For (E), focus on the center of the front face and the center of the right face. For the center of the front face any bar in the width direction and any bar in the height direction are excluded since the bars do not have a white cube in the middle. For the center of the right face any bar in the depth direction and any bar in the height direction are excluded because it would have the white color in the middle. Thus, the potential bars of both centers would have to go inside the big cube and each potential bar would contain the center of the big cube, but the center can only belong to one bar, not both.

22. (D) 6

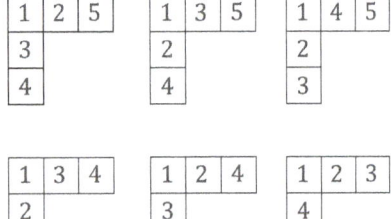

1 is the smallest number among 1, 2, 3, 4 and 5, so it can be written only in the upper left cell. 5 is the largest, so it can be written either in the rightmost cell or in the bottom cell. If 5 is written in the rightmost cell, then there are 3 options for writing all numbers. The other 3 options are obtained by switching the row and the column. Altogether, writing the numbers in the five cells can be done in 6 ways.

© Math Kangaroo in USA, NFP   130   www.mathkangaroo.org

SOLUTIONS 2017

23. (D) 13

R stands for each kangaroo looking to the right and L stands for each kangaroo looking to the left, so RRR L RR LL is the initial configuration. Let's focus on the kangaroos facing left. After 3 jumps of the first L to the left, the configuration changes to L RRRRR LL. The next L makes 5 jumps to the left, so the configuration becomes LL RRRRR L. The last L makes 5 jumps to the left, so the final configuration is LLL RRRRR and no more jumps are possible.
$3 + 5 + 5 = 13$ exchanges were made.

24. (C) 3

Look at numbers 2, 3, 12, 18 and a different fifth number. We have to multiply each of them either by 2 or by 3 (our choice). For example, the first four products can be 2×3, 3×2, 12×3, 18×2 and there are only two different results among the products: 6 and 36.
If the fifth number is multiplied by 2, then 3×2, 18×2, and the fifth $\times 2$ are different numbers.
If the fifth number is multiplied by 3, then 2×3, 12×3, and the fifth $\times 3$ are different numbers.
In each case, among the five products there are only three different results.
For any choice of 5 different numbers where we multiply each number either by 2 or by 3, we want to show that among the five products there are at least three different results.

Case 1. If you multiply five different numbers by 2, you have five different results. Same is true if you multiply five different numbers by 3.

Case 2. If you multiply four different numbers by 2 and another number by 3, then you have at least four different results. Same is true if you switch 2 and 3.

Case 3. If you multiply three different numbers by 2 and the other two numbers by 3, then you have at least three different results. Same is true if you switch 2 and 3.

These are all possible cases, so the smallest number of different results is 3.

25. (C) one triangle, one square

There are 4 gray and 8 white triangles, so at each edge of the big square you need 1 gray and 2 white triangles. The order along each edge should become gray, white, and white. This can be done by exchanging the white bottom triangle of the right edge with the gray middle triangle of the bottom edge. There are 8 white and 4 small gray squares. Only one exchange of the squares is sufficient, as shown, so the 4 center tiles of the big square are totally white and the other small squares along each edge appear in the order white and gray.

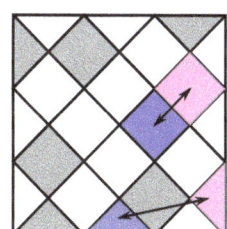

26. (C) 9

The number of green marbles cannot exceed 4. Otherwise, we could have 5 marbles with no red marble among them. Similarly, the number of red marbles cannot exceed 5. Otherwise, we could have 6 marbles with no green marble among them. Thus, the largest number of marbles in the bag is 9: 4 green and 5 red marbles.

SOLUTIONS 2017

27. (D) Beata, Celina, Ala
Beata picked her numbers, among them 45 = 3 × 3 × 5, before Celina picked her numbers. Otherwise, Celina would have picked 45 as a multiple of 5.
Celina picked her numbers, among them 20 = 5 × 2 × 2, before Ala picked her numbers. Otherwise, Ala would have picked 20 as a multiple of 2.
Therefore, the girls approached the basket in the order: Beata, Celina, Ala.
Among the eight numbers 20, 24, 25, 32, 33, 35, 45, and 52, only 24, 33, and 45 are multiples of 3 and Beata took all of them, so the numbers left are 20, 25, 32, 35, and 52. Among them only 20, 25, and 35 are multiples of 5 and Celina took all of them. The numbers left are 32 and 52. Both of the numbers are even, so Ala took these last two numbers.

28. (D) 7
In the diagram, "O" stands for an *odd* number and "E" stands for an *even* number. *odd* + *odd* is always *even*, *odd* + *even* and *even* + *odd* are always *odd*, and *even* + *even* is always *even*. The diagram to the right shows that John can write 7 odd numbers.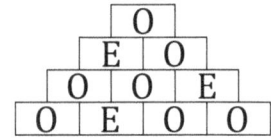
If the bottom row contained 4 odd numbers, then all rows above it would contain only even numbers, so the total of odd numbers would be only 4. Thus, 3 odd numbers in the bottom row is an optimal number. If the row directly above the bottom row contained 3 odd numbers, then the two rows above it would contain only even numbers, so the total of odd numbers would not exceed 3 + 3 + 0 + 0 = 6 odd numbers. The top two rows together cannot have more than 2 odd numbers, so the total number of odd numbers for the whole diagram cannot exceed 3 + 2 + 2 = 7 numbers.

29. (E) 48
Only two countries do not share a common border (the western country and the south-eastern country). We can pick any of the 4 colors for the biggest country and then any of the 3 remaining colors for the small country between the two countries that do not share a border. After that, two colors remain and we can use them without further restrictions to color in the two countries. Therefore, the map can be colored in 4 × 3 × 2 × 2 = 48 ways.

SOLUTIONS 2017

30. **(C) 6**

    6 lamps lit at the beginning is enough to eventually light all the lamps, as shown below. In the diagram, L stands for a lamp lit from the very beginning. The lamps light below and above each L. They are marked by 1. Above/below all 1s the new lamps are lit. They are marked by 2. Then lamps marked by 3 are lit, and so on. If less than six lamps are lit at the beginning, then there is a row that has no lit lamp at the start. One or more minutes later we want to light at least one lamp in the row. At this stage, the lamp to be lit has no neighboring lit lamps in its row and even if it has neighboring lit lamps above and below, the three cells (the cell of the lamp to be lit and the cells above and below it) have no common vertex, so the lamp to be lit will stay dark. Thus, at the beginning we need at least 6 lit lamps, one for each row and column, so that eventually all lamps will be lit.

5	4	3	2	1	L
4	3	2	1	L	1
3	2	1	L	1	2
2	1	L	1	2	3
1	L	1	2	3	4
L	1	2	3	4	5

Solutions for Year 2019

1. (B)

    Compared to the starting picture, (A) has longer nose, (C) has whiter ears, (D) has an inverted nose, and (E) does not have a triangular nose. Only (B) can be Carrie's finished drawing.

2. (C)

    Answer (A) represents $5 + 1 = 6$, (B) represents $5 + 5 + 1 = 11$, (C) represents $5 + 5 + 5 + 2 = 17$, (D) represents $5 + 5 + 5 + 4 = 19$, and (E) represents $5 + 5 + 2 = 12$. Thus, Mayan people wrote 17 using 3 bars and 2 dots.

3. (C)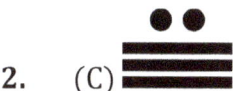

    The current time is 20:19 (8:19 PM) and the options for the next time displays are: (A) 20:91; (B) 9:21; (C) 21:09; (D) 9:12; and (E) 2:19. Valid minute displays are from 00 to 59, so (A) is not a valid display. 02:19, 09:12, and 09:21 are valid displays but they represent different times after midnight. 21:09 is a valid time display that occurs after 20:19 but before midnight, so the next time after 20:19 is (C).

4. **(E) 1**

    The total number of children in the kindergarten class is $14 + 12 = 26$, so 13 children go for a walk. At least one of them must be a girl since all 12 boys can go for a walk.

© Math Kangaroo in USA, NFP   133   www.mathkangaroo.org

SOLUTIONS 2019

5. (E)

The sums of dots on opposite faces need to be equal to 7. The pictures do not show opposite faces but only adjacent faces, so if two of the shown faces of a particular die have the sum of their dots equal to 7, then the die is not an ordinary die. This is the case for (A), (B), (C), and (D) since $2 + 5 = 7$, $4 + 3 = 7$, $3 + 4 = 7$, and $1 + 6 = 7$, respectively. Only (E) could be an ordinary die.

6. (D) 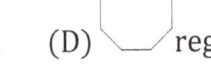 regular octagon

 A triangle is shown in green, a square is shown in yellow, a regular hexagon is shown in red, and a regular dodecagon (12-gon) is shown in blue. There is no regular octagon in the design.

7. (D) 8
At each of the four corners we can color one square of the size 2×2.

 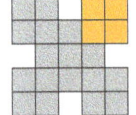  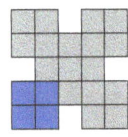

The area for the other 2×2 squares that can be colored is the 3×3 square with the borders in bold. We can color four 2×2 squares inside the 3×3 square, as shown below.

  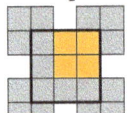  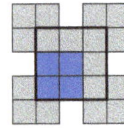

Altogether, there are eight 2×2 squares that can be colored.

8. (C) 20
The sum of any three odd numbers is always an odd number, so 20 cannot be the sum.
The other options can happen since the 6 smallest odd natural numbers are 1, 3, 5, 7, 9, and 11.
For example, (A) $21 = 5 + 7 + 9$, (B) $3 = 1 + 1 + 1$, (D) $19 = 7 + 7 + 5$, (E) $29 = 9 + 9 + 11$.

9. (B) 12
In two years the sum of the ages of kangaroos goes from 36 to 60 years, so the age increase is 24 years during the two-year period. Hence, the yearly increase is half of 24, which is 12 years. 12 kangaroos are needed for the sum of the ages to increase by 12 years in one year.

SOLUTIONS 2019

10. (A)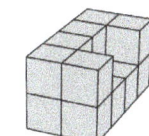

For each solid we count the number of unit squares of the surface. The numbers are: 42 for (A), 38 for (B), and 40 for the other three solids.
Here is another way to solve the problem:
When we look at each solid from below, we count $4 \times 2 = 8$ unit squares. The same is true when we look directly from the left, from the top, and from the right. When we look at (B) from the front or from the back we count only 3 unit squares in each case. For any other solid these numbers are 4 from the front and 4 from the back. Solid (A) is the only solid that has unit squares as part of the surface area that are not visible directly from any of the six directions. It has 2 such interior squares. Therefore, solid (A) needs the most paint.

11. (C) 9

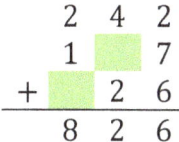 $3 + 7 + 6 = 16$, so we carry 1 to the tens column.

The missing digit in the tens column must be 5, since $1 + 4 + 5 + 2 = 12$, and this is the only way we can get a number ending in 2 for this column. Carry 1 to the hundreds column.

The missing digit in the hundreds column must be 4 since $1 + 2 + 1 + 4 = 8$.

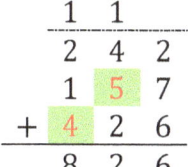 The sum of the missing digits is $5 + 4 = 9$.

© Math Kangaroo in USA, NFP   www.mathkangaroo.org

SOLUTIONS 2019

12. (C) 3

During 12 normal days Riri would eat $5+5+5+5+5+5+5+5+5+5+5+5 = 60$ spiders but there were only 9 days to eat all the spiders.
The above sum is equal to $5+5+5+5+5+5+(5+5)+(5+5)+(5+5)$, which is $5+5+5+5+5+5+10+10+10$ and it consists of exactly 9 numbers, one number for each day. 5s correspond to the days when Riri ate her usual number of spiders and 10s correspond to the days when she was very hungry. Therefore, Riri was very hungry on 3 days.

13. (C)

Below, we show how we can form figures (A), (B), (D), and (E).

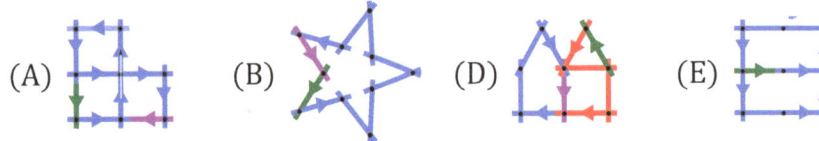

In each of the above cases, the green arrow is the first step and the purple arrow is the last step in our construction. For (D), the green arrow is followed by the red arrows and then by the blue arrows.

Let's look at figure (C). Pick any point as the starting point. Then, pick any blue point that is different than the starting point. When we reach this selected blue point the first time, we have to leave it but after that we have to return to this blue point again since it belongs to 3 parts of the folding yardstick. After the return we have to stop there since there is no additional part of the yardstick connected to it, which makes it the ending point. There are at least 3 blue points that are not the starting point, so each of them must be the ending point. This is impossible since every way of folding the yardstick has only one ending point.

14. (B)

Let's compute what fraction of all small squares is black in each answer option.

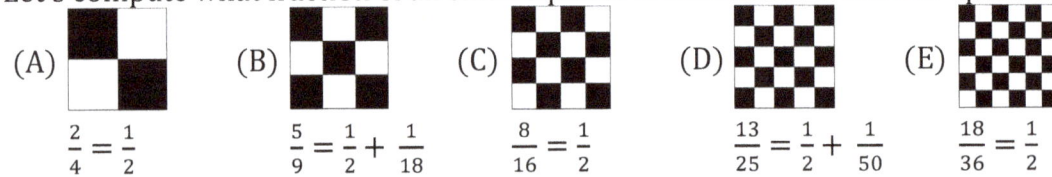

$\frac{2}{4} = \frac{1}{2}$ $\frac{5}{9} = \frac{1}{2} + \frac{1}{18}$ $\frac{8}{16} = \frac{1}{2}$ $\frac{13}{25} = \frac{1}{2} + \frac{1}{50}$ $\frac{18}{36} = \frac{1}{2}$

The largest fraction is $\frac{5}{9}$ for square (B), so square (B) has the largest black area.

SOLUTIONS 2019

15. (A) 15 m

The angles of the big triangle are equal to 60° since they are angles of equilateral triangles, so the big triangle is also an equilateral triangle. Any two equilateral triangles that share one side are equal, so the triangles marked with purple stars in the picture are equal to the triangle marked with the red star, and all sides of these three triangles are identical. By the same argument, all sides of the four small triangles (one shaded, one with a red triangular mark, and two with blue triangular marks) are equal to 1 m. Two sides of small triangles with blue triangular marks form one side of the triangle marked with a purple star. Thus, all sides of the triangles marked with stars are equal to 2 m. Therefore, the bottom side of the big triangle is equal to 1 m + 2 m + 2 m = 5 m and the perimeter of the big (equilateral) triangle is 3 × 5 m = 15 m.

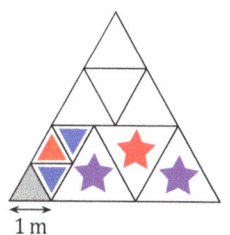

16. (C) 9

After two transformations there are 10 dogs, 10 cats, and 10 mice. 5 of these 10 mice were cats before the second transformation, so after the first transformation there were 10 dogs, 15 cats, and 5 mice. 6 of these 15 cats were dogs before the first transformation, so there were 16 dogs, 9 cats, and 5 mice at the beginning. Thus, at the beginning there were 9 cats.

17. (B) 11 cm

The dimensions which we need to calculate the heights of the towers are 1 cm for horizontal blocks and 2 cm for vertical blocks.

Using v for vertical and h for horizontal, the consecutive towers can be described, level by level, as 2v1h; 3v2h1v; 4v3h2v1h; and 5v4h3v2h1v. The numbers identify how many blocks are on a particular level. For example,

the 1st tower consists of 2 + 1 = 3 blocks and its height is 2 cm + 1cm = 3 cm,
the 2nd tower consists of 3 + 2 + 1 = 6 blocks and its height is 2 cm + 1 cm + 2 cm = 5 cm,
the 3rd tower consists of 4 + 3 + 2 + 1 = 10 blocks and its height is
2 cm + 1 cm + 2 cm + 1 cm = 6 cm,
and the 4th tower consists of 5 + 4 + 3 + 2 + 1 = 15 blocks and its height is
2 cm + 1 cm + 2 cm + 1 cm + 2 cm = 8 cm.
Following this pattern, the 5th tower consists of 6 + 5 + 4 + 3 + 2 + 1 = 21 blocks and its height is 2 cm + 1 cm + 2 cm + 1 cm + 2 cm + 1 cm = 9 cm;
the 6th tower consists of 7 + 6 + 5 + 4 + 3 + 2 + 1 = 28 blocks and its height is
2 cm + 1 cm + 2 cm + 1 cm + 2 cm + 1 cm + 2 cm = 11 cm.
Hence, the tower built from 28 blocks is 11 cm high.

SOLUTIONS 2019

18. (C) 9

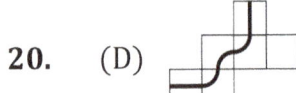

For the big square the green lines are the folding lines. After folding the big square twice, the small squares with the same letter will be folded on top of each other and the big square will be reduced to its lower right quadrant. The horizontal red lines of the big square are reduced to the horizontal red segment and the vertical red lines are reduced to the vertical red segment of the quadrant. The red segments of the quadrant are then its cutting lines, so the red lines are the cutting lines of the big square. Hence, the big square will be cut into 9 pieces formed along the red lines. After the two cuts, Bridget got 9 pieces of paper.

19. (A) both Alex and Carl

Bob is not wearing a hat today. By the second rule ("If Bob doesn't wear a hat, then Carl wears a hat."), Carl must wear a hat. By the first rule ("If Alex doesn't wear a hat, then Bob wears a hat."), Alex must wear a hat to make sure that Bob is not wearing a hat. Therefore, both Alex and Carl are wearing hats.

20. (D)

For each net below, the sides of small squares that are not shared with another small square are labeled from a to g in such a way that the sides with the same label form one edge of the cube obtained by folding the net.

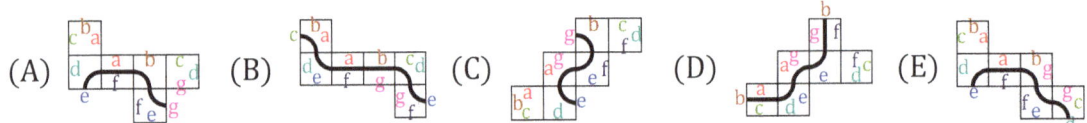

Net (D) shows the midpoints of two sides with the same label "b" as the endpoints of the drawn line, so net (D) becomes a cube for which the endpoints of the line drawn on it are connected (the line is **closed**). Each of the other nets shows different labels for the sides with the endpoints of the line, so only the net (D) represents the cube with the closed line.

21. (C) 41

The common product is a multiple of 5, 10, and 15. The least common multiple of these numbers is 30, and 6, 3, and 2 are complementary factors for 5, 10, and 15, respectively. The sum of these six numbers is $5 + 10 + 15 + 6 + 3 + 2 = 41$.
In general, if a positive integer P is the common product, then $\frac{P}{5}, \frac{P}{10}$, and $\frac{P}{15}$ are the numbers written on the faces opposite to the faces with 5, 10, and 15, respectively.
The sum of the six numbers is $5 + 10 + 15 + \frac{P}{5} + \frac{P}{10} + \frac{P}{15} = 30 + \frac{11 \times P}{30}$. It is a natural number only if the positive P is a multiple of 30, so the sum is the smallest if P is 30.
Thus, the smallest sum is 41.

SOLUTIONS 2019

22. (E) 90 g
Removing one black bead from each side of the balanced scales (on the right in the problem) shown in the top drawing keeps the scales balanced as shown right below it. Compare it to the other given scales (shown as 3rd from the top) with only black beads on its left side. Two black beads can be replaced by two white beads and the 6 g weight, so the last scale is also balanced. 30 g – 6 g = 24 g, so 3 white beads weigh 24 g. Thus, 2 white beads weigh $\frac{2}{3} \times 24g = 16$ g and 2 black beads weigh 6 g more, so 2 black beads weigh 22 g and 1 black bead weighs 11 g.
Originally, there were 6 black beads and 3 white beads on the two sets of scales. The 3 white beads weigh 24 g and the 6 black beads weigh 6 × 11 g = 66 g. The total weight of these nine beads is 24 g + 66 g, which is 90 g.

23. (D) My son Basil has 2 brothers.
The two statements, "My daughter Ann has 2 brothers," and, "My son Basil has 2 brothers," are contradictory statements. If Ann has 2 brothers, then Basil has only 1 brother. If Basil has 2 brothers, then Ann has 3 brothers. Thus, one of these two statements must be false, so the other three statements must be true since only one of the five statements is false. Hence, it is true that Basil has 3 sisters, Ann has 2 sisters, and there are 5 children altogether. 2 children must be boys, so the statement (D), "My son Basil has 2 brothers," must be false.

24. (C) 2
The 4th circle always contains a multiple of 3 as the result of multiplying by 3. Adding 2 to a multiple of 3 never results in a multiple of 3. Also, multiplying any non-multiple of 3 by 2 never results in a multiple of 3. Hence, the 4th circle always contains a multiple of 3, and the 5th and 6th circles never contain multiples of 3.
In terms of divisibility by 3, there are three types of integers: multiples of 3, 1 + a multiple of 3, and 2 + a multiple of 3. Among any three consecutive integers exactly one is a multiple of 3. In the first three circles there are always three consecutive integers, so exactly one of them is a multiple of 3. Therefore, among the six numbers in the circles exactly 2 are divisible by 3.

25. (B)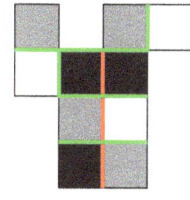

The cardboard cannot be folded along the middle vertical line (in red) to get any box. All the folding lines are shown in green. Keep the row with two black squares in front of you, fold back the row below it, and then the next row will fold to the back of this box. The two gray squares from the top of the cardboard will become the top face of the box, so the box looks like (A). Turn the box multiple times toward yourself to see (E), (C), and (D). You cannot get (B) since there are no two white squares that can form one face.

26. (B) 4
We can list Emily's cousins and call them C1, C2, C3, C4, C5, C6, C7, and C8. In the first photo we will have C1, C2, C3, C4, and C5. In the second photo we have C8, C7, and C6 additionally C5 and C4 will be in the selfie again. So far we have taken 2 selfies and each cousin has been in a photo once, except for C5 and C4. Our next selfie will have C1, C2, C3, C6, and C7. Now each cousin has been in a photo twice, except for cousin C8, so we have to just take 1 more selfie with C8 and 4 other cousins, giving us a total of 4 selfies.
Or: We can ignore Emily in counting the people in the pictures. To get 8 cousins in at least 2 photos each, with 5 cousins per photo, we need $(8 \times 2) \div 5 = 16 \div 5 = 3$ R 1 pictures taken. The "remainder 1" means that there needs to be a fourth photo because one of the cousins did not fit in the first three photos taken. This cousin will be in a picture with four other cousins, for whom this will be the third photo. If a fifth photo is taken, at least one of the cousins would have to be in four pictures, which is not the case. Thus, four is the number of photos.

27. (D) 26
By comparing parts of two identical pyramids of 15 cans, we notice that the missing cans in the first pyramid, in the four rows shown from bottom up are: 3, 8 & 2, 3, and 4. At the top of each pyramid there must be one more can, since without it the picture only shows 14 cans. $3 + 8 + 2 + 3 + 4$ is equal to 20, and Jette knocked down 6 cans for a total of 25 points, so there was a can
Jette's throw   Willi's throw

worth 5 points at the top of the pyramid. The points for the missing cans in both pyramids are shown. The second pyramid shows Willi's four missing numbers. Their sum is $5 + 9 + 4 + 8 = $ = 26, so Willi scores 26 points.

28. (A)
The current time is shown as , but the first digit of a time display in 24-hour format can only be 0, 1, or 2. The first digit of the current time cannot be 0 or 1 due to the horizontal middle segment. It must be 2. The two non-working segments in the first digit are shown in red here .
The next digit on a 24-hour clock can only be 0, 1, 2, 3, or 4. Only the digit 3 can be formed with the segments shown and with the upper right vertical segment that is not working. A similar analysis shows that the next two digits are 4 and 7, so the current time is 23:47. 3 hours and 45 minutes from now it will be 3:32.
This time is shown in (A) as as it actually is .

Here are the other times listed in the answers: (B) represents 00:32, (C) represents 01:23, and (D) represents the hour 01 and the minutes could be 35, 36, 38, or 39. (E) doesn't have a valid representation for the first two digits.

SOLUTIONS 2019

29. (D) $\frac{3}{4}$

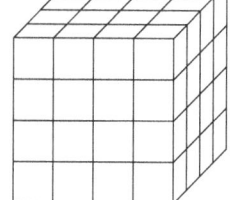

Any cube has 8 corners and 12 edges, so the large cube has 8 small cubes at the corners. Along each edge of the $4 \times 4 \times 4$ cube there are other 2 small cubes, so along 12 edges there are 24 such small cubes. $8 + 24 = 32$, which matches exactly the 32 small white cubes given. The rest of the small cubes are the 32 small black cubes.
To make the largest area of the surface white, we place all 32 small white cubes along all the 12 edges (which does include the corners) of the large cube. Each white cube at each of the 8 corners adds 8×3 white unit squares to the surface of the large cube and the other 24 white cubes add 24×2 white unit squares to the surface. There are $8 \times 3 + 24 \times 2 = 24 + 48 = 72$ unit squares of the surface which are white. The whole surface consists of $6 \times (4 \times 4) = 96$ unit squares, so $\frac{72}{96} = \frac{3}{4}$ of the surface is white.

30. (C) 14

At the start Zev has 4 white tokens and no red tokens, so his first step must be an exchange of 1 white into 4 red tokens. If he repeats this 3 more times, then after these 4 exchanges Zev has 16 red tokens and no white tokens. His next step must be an exchange of 1 red into 3 white tokens and at this moment he has 15 red tokens and 3 white tokens.
At this point 5 exchanges took place. We need 6 more exchanges to have the total of 11. Not all of the last 6 exchanges can be from 1 red into 3 white tokens since the number of red tokens would decrease from 15 to $15 - 6 = 9$ and the number of white tokens would increase from 3 to $3 + 6 \times 3 = 21$ for a total of $9 + 21 = 30$ tokens. However, 31 tokens are needed. Thus, after 5 initial exchanges we need just one of the following 6 exchanges to be from 1 white to 4 red tokens. The number of all tokens can be easily computed if the 1 white to 4 red tokens exchange is the 11th one.
So, after the initial 5 exchanges when Zev had 15 red and 3 white tokens, the next 5 exchanges will be from 1 red into 3 white tokens, decreasing the number of red tokens from 15 to $15 - 5 = 10$ and increasing the number of white tokens from 3 to $3 + 5 \times 3 = 18$. The 11th exchange will be from 1 white to 4 red tokens, decreasing the number of white tokens from 18 to $18 - 1 = 17$ and increasing the number of red tokens from 10 to $10 + 4 = 14$. After 11 exchanges there are $14 + 17 = 31$ tokens and 14 of them are red. Additionally, among 11 exchanges 5 are from white to red and 6 are from red to white.

Solutions for Year 2021

1. (D)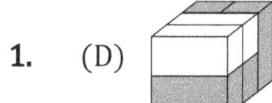

Among the given bricks, there are 4 shaded bricks and 2 white bricks. Only (D) has this number of shaded and white bricks (the other options all have 3 of each type of brick).

SOLUTIONS 2021

2. (A) 1
When a child is facing the back, their left hand is on the left, whereas if a child is facing the front, their left hand is on the right, from our perspective. For two children to be holding each other with their left hands, a front-facing child must be holding the hand on our right to a back-facing child. This only happens once in the chain of children, where the 5th and 6th children are holding hands.

3. (E)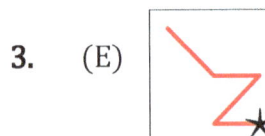
To create the largest number, we want the digits in the front of the number (i.e., the most significant digits) to be as large as possible. Therefore, we can narrow it down to choices (B) and (E) because they both start with 98. The next digit in (B) is 5 and the next digit in (E) is 6, so (E) is larger than (B).

4. (D) G
Since Sofie must take one letter from each box, she is forced to take N from box 3 because it is the only letter. Then, box 1 only has one option because N has already been selected, so we choose E from box 1. Similarly, box 5 only has one option because E is already chosen, so we take U from box 5. If we continue this logic, Sofie will have to choose G from box 4.

5. (B) 32
The only way to fit the puzzle pieces is the following:
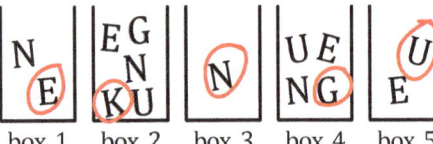
$12 + 20 = 32$

6. (C) 69
From the first level of the measuring tape to the second level, the numbers are increasing by $27 - 6 = 21$. Therefore, the number at the question mark should be $27 + 21 + 21 = 69$.
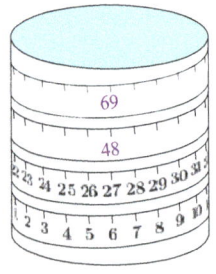

7. (B) B
For a figure to be able to leave through gate G, its maximum width at any row in the figure must be 3 blocks. The only figure that satisfies this is B. Once the first 2 rows of B exit gate G, it can be shifted to the left so that the last row can leave through gate G.

SOLUTIONS 2021

8. **(E) They will all be equally dark.**
   All the choices give a proportion of 1 part of green to 3 parts of white, so they are the same color.
   $$\frac{1}{3} = \frac{2}{6} = \frac{3}{9} = \frac{4}{12}$$

9. **(E) any of P, Q, or R**
   Folding any isosceles right triangle like R along its axis of symmetry will still give Mary the same triangle shape, no matter how many times it is folded. The square Q can be folded along its diagonal to create a right triangle like R and folded once more to maintain the triangle shape. And finally, the rectangle P, whose sides are of the ratio 1:2, can be folded by its vertical axis of symmetry to create a square and then folded once more, like Q, to form a triangle.

10. **(D)**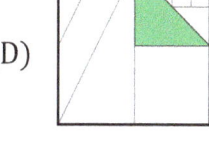

    Since all the line segments are drawn from midpoints, the square in the bottom right is 1/4 of the entire larger square. Half of this smaller square will have an area of 1/8 of the larger square, so the only correct option here is the green triangle.

11. **(B) 3444**
    Since the number has 10 digits, the smallest sum will be the sum of a 4-digit number and two 3-digit numbers. The 4-digit number we make must be as small as possible. We can minimize the 4-digit number by having 1 as its first digit. This separates the number into 502, 1972, and 970. The sum of these numbers is 502 + 1972 + 970 = 3444.
    Note: We may start with 502 and 197, so that the third number is 2970. Their sum is 3669. The answer to the question actually depends on the order of the digits in our original 10-digit number. For example, for the number 5201972970 we could use 52 + 01972 + 970 = 2,994, which is less than 3,000. For our original number 5021972970 we cannot use a five-digit number as one of the three numbers formed. If one of the numbers is an actual 5-digit number, then the sum is at least 10,000 (much more than 3669). If this five-digit number starts with 0, then for 5021972970 the sum is 5 + 02197 + 2970, which is more than 3669. Thus, to minimize the sum we cannot use five (or more) digits for any of the three numbers. We have to use four digits for at least one of them since 3 + 3 + 3 is less than 10. 2 + 4 + 4 = 4 + 2 + 4 = = 4 + 4 + 2 = 10, so our sums could be 50 + 2197 + 2970, 5021 + 97 + 2970, or 5021 + 9729 + 70. Each of the three sums is greater than 3669, so the only options left are 4 + 3 + 3, 3 + 4 + 3, and 3 + 3 + 4. The corresponding sums are 5021 + 972 + 970 = 6963, 502 + 1972 + 970 = 3444, and 502 + 197 + 2970 = 3699. Therefore, the smallest sum is 3444.

SOLUTIONS 2021

12. (B) 20 km
If we add the lengths of the described round trips from A, B, and C, we get the outer path (brown line) plus the inner path (red line). So, if from this we subtract the inner path, we get the length of the required outer path, namely 10 + 12 + 13 − 15 = 20 km.

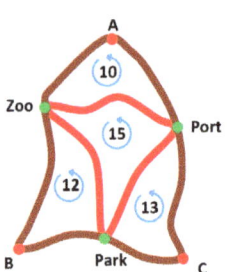

13. (D)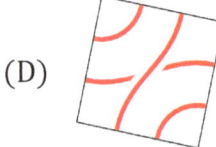

To complete the path between the two arrows, we want to connect the two lines coming out of the arrows. This can be done with either 2 horizontal lines in the center piece or small circular sector to connect the 2 lines on the left edge of the center piece. Since the pieces can be rotated, A and C both have a circular arc that will connect the appropriate lines, and B and E each have 2 sets of horizontal lines that will complete the path. Therefore, D is the only one that won't fit.

14. (B) 4
The sum of the visible numbers on the left hexagon is 3 + 6 + 1 + 2 = 12. As the sum of all the numbers on the hexagon is 30, the sum of the numbers at vertices A and B is 30 − 12 = 18. Similarly, on the right hexagon the visible numbers add up to 23, so the sum of the numbers at vertices C and D is 30 − 23 = 7. Now look at the central hexagon. The sum of the numbers at A, B, C, D, and the 1 at the top right vertex is 18 + 7 + 1 = 26. So, what remains is 30 − 26 = 4.

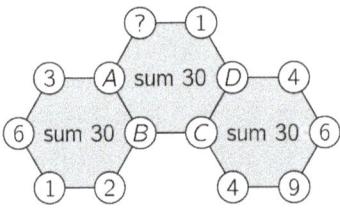

15. (C) 8 cm
We can apply the formula for the area of rectangle: width × height. From the combined area of the first two rectangles, we see that their height h = (12 + 18) ÷ 6 = 5 cm. So, CD = (18 + 22) ÷ 5 = 8 cm.

16. (A) A
The ball on the top of the pyramid is shown in all three side views. It has the letter A on it. The second ball with A on it has to appear on at least one of the side views. Since this is not the case, it must be the one with the question mark.

© Math Kangaroo in USA, NFP www.mathkangaroo.org

SOLUTIONS 2021

17. (E)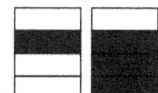

    Here, using "r" and "ℓ" correspond to either Ronja or Wanja placing their token in the right or left pile, respectively. The following are the sequence of moves that will fulfill each of the token placement choices.
    (A) R(ℓ), W(ℓ), R(r), W(ℓ), R(r), W(r), R(r), W(ℓ);
    (B) R(ℓ), W(r), R(ℓ), W(ℓ), R(r), W(r), R(r), W(ℓ);
    (C) R(ℓ), W(r), R(r), W(ℓ), R(r), W(ℓ), R(ℓ), W(r);
    (D) R(ℓ), W(r), R(r), W(ℓ), R(ℓ), W(r), R(ℓ), W(r);
    (E) Each game ends at the top of one pile. Ronja starts the game, so Wanja must end it. Therefore, at least one token at the top must be gray, so the players could not create the piles of (E).

18. (C)

    Comparing the first number of the original combination to the first number of this combination, we know that each number must have been rotated backwards twice in order for the number to go from 6 to 4. If we look at the second digit now, rotating twice backwards would result in a 1 for the second digit, and not 9, so 4906 cannot be the correct combination.

19. (D) Carl got as many pears as Luca got apples.
    The number of Carl's apples plus the number of Luca's apples equals 20. The number of Carl's apples plus the number of Carl's pears also equals 20. If Carl got p pears, then he got $20 - p$ apples because he selected 20 total fruit, and Luca got the remaining p apples. Hence, Carl got as many pears as Luca got apples.

20. (B)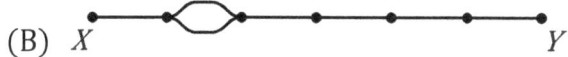
    Let the vertices from X and to Y be labeled as A, B, C, D, and E, respectively.

    Since the trip from X to Y and Y to X is split into 6 segments, we know that each segment from X to Y takes $180 \div 6 = 30$ min, and each segment from Y to X takes $60 \div 6 = 10$ min. The first train will take 30 min to go from X to A, whereas the other train will go from Y to C in the same 30 min. In the next 30 min, while the first train is traveling between A to B, the other train will travel from C to X, resulting in a collision on the path from A to B. Therefore, the double track should be there.

© Math Kangaroo in USA, NFP   www.mathkangaroo.org

SOLUTIONS 2021

21. (A) Ann and Bob
Since the table is round, each person has exactly 2 neighbors. Bob is not sitting next to Ann nor Dan. Therefore, his 2 neighbors must be Carina and Ed. Ed's other neighbor is Dan, so Ann is the last person to be placed, and she is between Dan and Carina. Therefore, Carina's neighbors are Ann and Bob.

22. (B) 8
If 100 pancakes require 25 eggs, 4 l of milk, 5 kg flour, and 1 kg butter, then 1 pancake requires 0.25 eggs, 0.04 l of milk, 0.05 kg flour, and 0.01 kg butter. Look at each ingredient separately. With 6 eggs, he can make $6 \div 0.25 = 600 \div 25 = 24$ pancakes; with 0.5 l of milk, he can make $0.5 \div 0.04 = 50 \div 4 = 12.5$ pancakes; with 0.4 kg flour, he can make $0.4 \div 0.05 = 40 \div 5 = 8$ pancakes; and with 0.2 kg of butter, he can make $0.2 \div 0.01 = 20 \div 1 = 20$ pancakes.
The limiting ingredient, therefore, is the flour, with which he can only make 8 pancakes.

23. (A)

The black tooth of the first gear moves 10 positions clockwise. So, the black tooth on each of the other gears also moves 10 positions, but counterclockwise for the middle gear and clockwise for the top one. Count 10 teeth away, in the appropriate directions, to get figure (A) as the answer.

SOLUTIONS 2021

24. (C) peach
The pictures below show the three original statements.

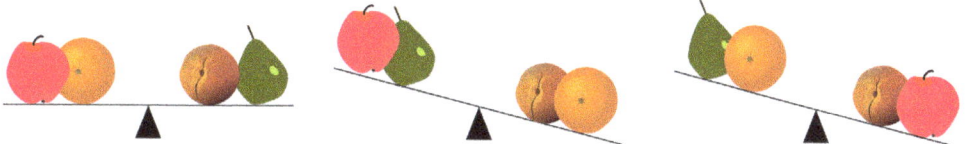

Since the two sides of the first balance are the same, we can add each pair to either side of the one of the other balances. Starting with the second balance, we can get:

Removing the pear and orange from both sides, we see that two apples are lighter than two peaches, so a peach is heavier than an apple.

Using the same balance but adding the fruit from the first scale on opposite sides, we get:

Removing an orange and an apple from either side, we see that two pears are lighter than two oranges, so an orange is heavier than a pear.

Now we can try adding the fruit from the first balance to the second balance. Keeping them on the same sides, we get:

Removing a pear and an apple from both sides shows that two oranges are lighter than two peaches, so a peach is heavier than an orange.
We now know that a peach is heavier than an orange as well as an apple, and an orange is heavier than a pear. This is enough information to conclude that a peach is the heaviest of the fruits.

SOLUTIONS 2021

25. (E) 21

To make sure the design is symmetrical over the vertical axis, we will need to color 3 additional squares. Then to make the design symmetrical over the horizontal axis, the 6 squares that are already there will need to be reflected onto the bottom, so we will need to color 6 additional squares. Then, to create axes of symmetry over both diagonals of the square, we will need to add the same 6-block design to both the left and right side of the square. The total squares that need to be colored will be $3 + 6 + 6 + 6 = 21$.

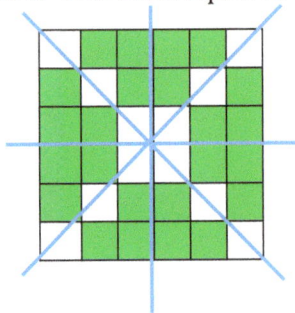

26. (C) 13

If the statement of pirate 2 and pirate 3 about the coins were false, then their statements about the diamonds would both be true. But, this cannot happen as they claim different numbers of diamonds. So, the friend had 7 coins. Pirate 1 is lying about the number of coins, so he is telling the truth about the number of diamonds, which is 6. The total number of coins and diamonds is $6 + 7 = 13$.

27. (D) 10

If S, M, and L are the volumes of the small, medium, and large bottles, respectively, then we can set up the following equations: $3L + 4S = 64$; $3S + 2L + 2M = 64$; $6S + 4M = 64$. Dividing both sides of the equation by 2, the equation $6S + 4M = 64$ becomes $3S + 2M = 32$.
$3S + 2M + 2L = 64$, so $32 + 2L = 64$. $L = (64 - 32) \div 2 = 16$. $3L + 4S = 64$, so we can plug in $L = 16$ and get $S = (64 - 3 \times 16) \div 4 = 4$. Plugging in $S = 4$ into the equation $6S + 4M = 64$ gives us $6 \times 4 + 4M = 64$, so $M = (64 - 6 \times 4) \div 4 = 10$. The final solution is $L = 16$, $M = 10$, and $S = 4$.

28. (B) 62

Each diagonal has 7 cubes with a line, so each face has $7 + 7 - 1 = 13$ (the "$- 1$" is for double counting the central square on the face). So, all 6 faces of the cube have 13×6 cubes with lines. However, we are triple counting the small cubes on each of the 8 corners because they appear on 3 different faces, and we only want to count them once, so we will subtract each of those cubes twice from the total. Therefore, the total is $13 \times 6 - 2 \times 8 = 62$.

SOLUTIONS 2021

29. (B) 3

The actual sum of all the numbers on the tokens of the elves and trolls is $1 + 2 + \ldots + 10 = 55$. However, the sum of the group's answers was 36, so some trolls answered with a smaller number than is on their token. To find the smallest number of lying trolls that could be in the group, we will want to start with the largest amount by which each troll is lying. If one troll said 1 but their token had a 10, and another troll answered 1 but their token had 9, then that will bring the sum to $55 - 9 - 8 = 38$, which is not low enough. Therefore, we know there must at least be one other troll that is lying, meaning there are at least 3 trolls. An example of this is if the elves have the tokens 1, 2, 3, 4, 5, 6, and 7, and the trolls have the tokens 8, 9, and 10, but they each say 4, 1, and 1, respectively, so the total sum will be 36.

30. (E)

There are only four possible options of cards that can make up the left edge of the rectangle. They are the following:

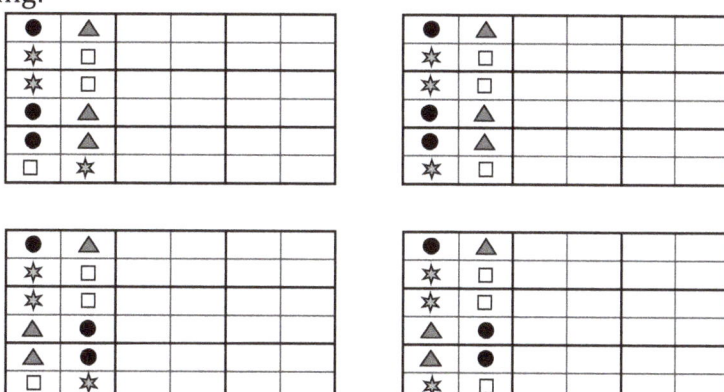

Notice that the circle and the triangle will always be in the same row on the card, and the star and the square will be in the same row. So, the following columns don't occur:

Therefore, (E) was definitely not used to form the rectangle.
Below are examples of how the other cards can be used:

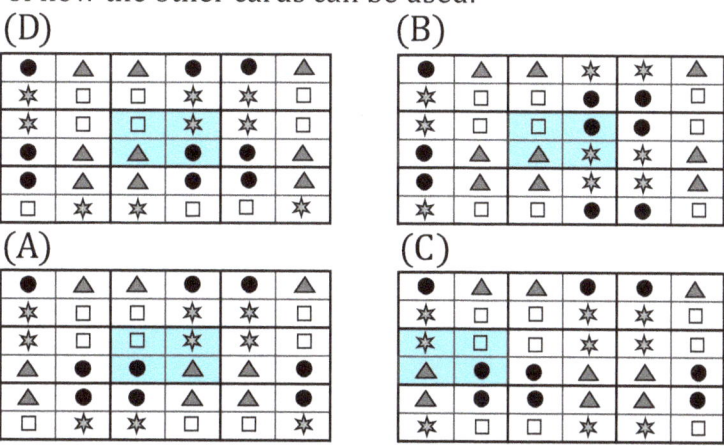

Part III: Answer Keys

ANSWER KEYS

	1999	2001	2003	2005	2007	2009
1	B	B	C	E	D	D
2	D	D	A	C	D	B
3	B	C	A	B	D	B
4	D	B	C	D	C	C
5	B	E	B	D	B	A
6	D	C	A	D	B	B
7	D	D	E	D	A	D
8	B	E	D	A	D	C
9	B	C	A	E	C	C
10	A	A	C	C	A	E
11	D	D	C	C	D	C
12	D	A	B	D	C	C
13	B	C	D	C	C	D
14	E	C	C	E	B	E
15	C	B	B	B	B	D
16	A	D	D	B	C	C
17	C	C	B	B	B	B
18	A	D	B	E	C	E
19	D	E	C	C	D	D
20	A	B	A	E	B	C
21	D	A	C	E	D	C
22	C	B	D	D	B	B
23	D	A	D	D	D	C
24	D	C	B	D	E	E
25	E	E	D	D	E	C
26	B	D	E	C	B	C
27	E	A	A	D	C	E
28	B	C	D	D	D	D
29	B	C	C	B	A	A
30	E	D	B	D	A	A

ANSWER KEYS

	2011	2013	2015	2017	2019	2021
1	C	E	B	B	B	D
2	B	C	C	C	C	A
3	D	C	A	E	C	E
4	A	B	A	D	E	D
5	E	E	B	B	E	B
6	B	B	E	A	D	C
7	B	B	D	E	D	B
8	D	E	A	C	C	E
9	E	C	A	B	B	E
10	D	C	D	D	A	D
11	D	C	E	D	C	B
12	B	C	C	C	C	B
13	E	D	C	C	C	D
14	C	B	C	C	B	B
15	B	E	C	D	A	C
16	D	B	B	E	C	A
17	B	D	C	E	B	E
18	C	A	E	D	C	C
19	D	C	B	B	A	D
20	C	D	E	C	D	B
21	E	A	C	A	C	A
22	E	D	C	D	E	B
23	C	B	D	D	D	A
24	D	A	C	C	C	C
25	B	D	B	C	B	E
26	A	D	C	C	B	C
27	E	B	B	D	D	D
28	D	B	E	D	A	B
29	C	D	E	E	D	B
30	A	B	D	C	C	E

www.ingramcontent.com/pod-product-compliance
Lightning Source LLC
Chambersburg PA
CBHW041411300426
44114CB00028B/2981